Agriculture Issues and Policies

nova
science publishers
www.novapublishers.com

Agriculture Issues and Policies

Pomegranate: For Horticulture Students and Farmers
Arkalgud Nanjundaiah Ganeshamurthy, PhD (Editor)
2023. ISBN: 979-8-88697-812-4 (eBook)

Research Advancements in Organic Farming
Jatindra Nath Bhakta, PhD (Editor)
Sukanta Rana, PhD (Editor)
2023. ISBN: 979-8-88697-519-2 (Hardcover)
2023. ISBN: 979-8-88697-580-2 (eBook)

Legumes: Nutritional Value, Health Benefits and Management
Phetole Mangena, PhD (Editor)
2023. ISBN: 979-8-88697-558-1 (Softcover)
2023. ISBN: 979-8-88697-583-3 (eBook)

Jute: Cultivation, Properties and Uses
Matthieu Issa (Editor)
2022. ISBN: 979-8-88697-490-4 (Softcover)
2022. ISBN: 979-8-88697-505-5 (eBook)

Pistachios: Cultivation, Production and Consumption
Shaziya Haseeb Siddiqui, PhD (Editor)
Shoaib Alam Siddiqui, PhD (Editor)
2022. ISBN: 978-1-68507-949-9 (Hardcover)
2022. ISBN: 979-8-88697-063-0 (eBook)

More information about this series can be found at
https://novapublishers.com/product-category/series/agriculture-issues-and-policies/

Rodrigo Nogueira de Sousa
Thais Nascimento Pessoa
and Kassio Ferreira Mendes
Editors

Agricultural Soils

Agronomic Management

nova
science publishers
www.novapublishers.com

https://doi.org/10.52305/VXLZ8058

NOTICE TO THE READER

Library of Congress Cataloging-in-Publication Data

Names: Sousa, Rodrigo Nogueira de, editor. | Pessoa, Thais Nascimento, editor. | Mendes, Kassio Ferreira, editor.
Title: Agricultural soils : agronomic management / Rodrigo Nogueira de Sousa, Thais Nascimento Pessoa, Kassio Ferreira Mendes.
Other titles: Agriculture issues and policies series
Description: New York : Nova Science Publishers, [2023] | Series: Agriculture issues and policies | Includes bibliographical references and index.
Identifiers: LCCN 2023033882 (print) | LCCN 2023033883 (ebook) | ISBN 9798891130005 (hardcover) | ISBN 9798891130692 (adobe pdf)
Subjects: LCSH: Soil science. | Soil management. | Soil conservation.
Classification: LCC S596.7 .A35 2023 (print) | LCC S596.7 (ebook) | DDC 631.4--dc23/eng/20230825
LC record available at https://lccn.loc.gov/2023033882
LC ebook record available at https://lccn.loc.gov/2023033883

Published by Nova Science Publishers, Inc. † New York

Contents

Preface

This book is a comprehensive and up-to-date reference work that provides a detailed exploration of the key issues related to soil fertility, management, and conservation in agricultural settings. The book covers a range of topics including nutrient management, soil aggregation, soil organic matter, fertilizers, soil microbiology, and soil health. It provides insights into the latest research and practical strategies for improving soil fertility, productivity, and sustainability. This book is an essential resource for researchers, agronomists, soil scientists, and students interested in the field of agricultural soil management. This book is written for a diverse range of readers who are interested in the field of agricultural soil management, including professionals, academics, and students. It will be of particular interest to agronomists, soil scientists, environmental scientists, researchers, and professionals working in agriculture, horticulture, and land management. In addition to professionals, this book may also appeal to non-professionals in industries such as agriculture, food production, and environmental sustainability. It could also be useful to government agencies, NGOs, and policy-makers who are interested in promoting sustainable agriculture and soil conservation practices.

Chapter 1 - Phosphorus (P) is the highly sought-after element in tropical soil-based crop systems, primarily attributed to its inherently low bioavailability and limited occurrence in natural deposits. Given the scarcity and finite nature of these deposits, extensive research has been conducted on harnessing the potential of phosphorus solubilizing microorganisms (PSM) to enhance phosphorus availability to plants. These microorganisms solubilize phosphorus present in soils, converting it into plant-assimilable forms. The objective of this chapter is to elucidate the role of PSM in the solubilization of insoluble phosphate and the availability of phosphorus to plants. Furthermore, the chapter delved into the methodologies employed for the selection and isolation of phosphorus-solubilizing microorganisms (PSM). In-depth discussions ensued regarding the diverse applications of PSM in agriculture, with a particular focus on augmenting food production and remediating contaminated soils. Within this context, the identification of microorganisms proficient in solubilizing inorganic phosphorus, along with their associated mechanisms such as organic acid secretion, H+ proton extrusion, and siderophore production, was thoroughly examined. Notably, enzymatic processes were found to be the primary means by which organic phosphorus is mineralized. It became evident that harnessing the potential of PSM holds immense promise for facilitating the enhanced accessibility of phosphorus to plants, thereby bolstering agricultural efficiency and productivity.

Chapter 2 - Reduction of fertility is one of the most important problems of the agricultural lands in today's world. This situation is mostly due to the widespread of conventional agriculture to feed the ever-growing population of the world. As a result, unconscious and excessive utilization of fertilizers to obtain the highest yield from unit area led to unfertile and polluted soils. Fertilization performed without proper evaluation of the nutrient element capacity of the soils can cause the micronutrient elements to reach a toxic level and research showed that together with excess phosphorus fertilizer utilization cadmium heavy metal levels have raised. Another source of pollution is the reduction of organic matter in soils below the sufficiency level due to conventional agriculture applications. Organic matter which affects productivity capacity of soils resulted from inaccurate fertilizer applications to eliminate the deficiencies of micronutrient elements led to toxic levels of some micro nutrient elements. The most economical and easy method to apply to remediate the pollution based on fertilizer utilization in the agricultural lands is phytoextraction. This chapter discusses how to form some pollutants due to fertilizers and solutions to this problem.

Chapter 3 - Mineral fertilizers are readily soluble and available nutrients in the soil solution for plants uptake. These sources are necessary for food production to remain constant. However, the manufacturing process of conventional nutrient sources is surrounded by environmental problems. Fertilizers are still produced on a large scale, but, long-term mineral extraction can lead to a shortage of nutrient sources. Soil, in its complexity and heterogeneity, is the basis of agriculture. In addition to the problems associated with the production offertilizers, their use on soil can lead to low fertility over time by affecting its organic content. The physical, chemical, and biological quality of the soil must be maintained tosustain production. The biological attributes as organic matter and microorganisms, when affected, also affect chemical and physical attributes. Organic matrix fertilizers can be the key to increasingly sustainable agriculture and producing quality agricultural commodities. The disadvantage of organic fertilizers is the need to apply high doses to meet crop requirements. However, when organic fertilizers are combined with minerals, know as organo-mineral fertilizers, the amount of fertilizer needed to be applied is reduced, making application advantageous. Thus, the authors'objective in this chapter is to present the potential environmental problems of NPK fertilizer production and how the use of organic residues can be beneficial by providing an environmentally sounddestination for the residues and reducing dependence on mineral fertilizer imports from other countries.

Chapter 4 - Soil structure has as its main component the soil aggregate, which is composed primarily of particles of sand, silt, and clay, bound together by inorganic and organic materials at various scales. However, the concepts and models explaining how soil aggregates form and stabilize have evolved over the years. Nowadays, it is understood that microaggregates are primally formed from the disruption of macroaggregates. Nevertheless, free microaggregates can feed back into the system to form new macroaggregates. During this process, physical-chemical interactions such as flocculation, organic bonds, and inorganic cementing agents such as iron and aluminum oxyhydroxides give rise to soil microaggregates. The main methods utilized to evaluate the stability of the aggregates can be grouped by (i) the amount of clay dispersed (WDC); (ii) methods that evaluate soil aggregates by size classes or indices such as mean weight diameter (MWD) and geometric mean diameter (GMD) or (iii) by evaluating the tensile strength (TS). Organic factors are also undoubtedly extremely relevant to the formation and stabilization of soil aggregates. Soil ecosystem engineers as earthworms and termites mediate the formation of soil structure and aggregation. Earthworms can act as catalysts for the decomposition of soil organic matter by moving the soil matrix during the burrowing process. Moreover, earthworms produce mucus, and their guts and casting process mix soil, organic particles, and binding agents in the same process. On the other hand, termites are known to increase the formation of soil organic matter and the activity of microorganisms, as well as produce casts and mucus as binders for soil aggregation formation and stabilization. Furthermore, soil organic matter and the microorganisms, which are ubiquitous in most soils act mainly in the formation of macroaggregates and, consequently, influence the structure and porosity of the soil. The roots' action is due to the exudation of extracellular polymeric substances (EPS) and to the physical association of individual particles. Microorganisms act in a similar manner. Bacteria can secrete substances onto the colloidal surface that acts as a binding agent, or fungal hyphae can physically bind particles together. Soil aggregation is one of the parameters used to assess Soil Physical Quality (SPQ), an important component used to assess Soil Quality (SQ). In this way, it is possible to evaluate different land uses and soil management systems in terms of their quality. Soil aggregation is relatively sensitive to soil management because of the wide range of processes and mechanisms that are involved in the formation and stabilization of soil aggregation.

Chapter 5 - Soil organic matter (SOM) is a pivotal component for sustainable agriculture in any environment. SOM plays a crucial role in soil quality by providing positive effects on soil chemical, physical and biological properties besides increasing crop productivity. Furthermore, SOM is a component of the global carbon cycle, and changes in its content can mitigate or intensify the emission of greenhouse gases. High SOM content can ensure food security for a growing world population while reducing the adverse effects of climate change. Therefore, understanding SOM dynamics is decisive while creating strategies that enhance its accumulation and permanence in the soil. In general, these strategies should be based on increasing inputs and decreasing losses of SOM. Several natural (climate and soil) and anthropogenic (management practices) factors, in a dynamic interaction, drive the accumulation of SOM. The main climate factors are temperature and precipitation, which can be associated with net primary productivity and SOM decomposition rates. Soil factors are mainly texture, mineralogy, and aggregation, having the most effect on SOM stabilization processes. Finally, several management practices can be used as a strategy for SOM accumulation, including smart land use, organic amendments, biochar, mineral fertilization, liming, cover crops, and no-till. Worthwhile highlight that management practices can be changed and improved, contrary to climate and soil, whose characteristics are immutable in the short term. Thus, this chapter emphasized and discussed management practices with more emphasis.

Chapter 6 - The biological activity of soils is one of the most important parameters in sustainable agriculture and is crucial for the transformation of plant nutrients into extractable forms. Bacteria are the most important group of microorganisms that contribute to the vitality of agricultural soils. Bacteria, the largest group of microorganisms at neutral pH, increase the solubility and extractability of plant nutrients. They also help to improve the physical and chemical properties of the soil and create an environment suitable for biological production. However, due to some agricultural systems, the biological properties of the soil decrease over time. Bio-fertilizers step into the field to eliminate this problem. Bio-fertilizers are environmentally friendly organic agricultural inputs. Enrichment of agricultural soils with nitrogen results in enrichment of soils with phosphorus due to increased solubility of phosphate compounds and also stimulates the secretion of some chemicals that promote plant development. Biofertilizers also reduce the use of chemical fertilizers and protect against pathogenic fungi. Rhizobium, Azosipirillum, Azotobacter, Bacillus, Pseudomonas are among the most important groups of organisms. Nevertheless, bioremediation becomes prominent in the natural alleviation of micro-nutrient toxicity and especially heavy metal pollution resulting from industry, traffic, fertilizers, pesticides, municipal wastes, etc. This chapter focuses on the relationship of biological characteristics of agricultural lands with the environment and solutions.

Chapter 7 - Rhizosphere is defined as the volume of soil that surrounds and influences plant roots. It is considered the most active portion of the soil and depends on three factors: plant genotype, soil (or substrate) properties, and microbial population. Processes occurring in the rhizosphere lead to changes in soil properties. For this reason, some research refers to them as "rhizospheric effects." Roots release molecules or other substances in a process called rhizodeposition. Among the many substances released by plant roots are root exudates and mucilage. Root exudates and other substances are rich in carbon and are closely related to the increase in microbial activity and selection of the microbial population in the rhizosphere. Mucilage plays an important role in hydraulic changes in this environment. Substances released by roots cause chemical changes, and together with biological changes, lead to many important biogeochemical processes in the rhizosphere. They are also responsible for soil physical changes in the soil of the rhizosphere. Recent studies indicate that we can use the knowledge of changes in the rhizosphere to avoid stress conditions and improve plant growth and health. Understanding these changes for different cropping systems and in different environments will help in the selection and development of management practices that will increase plant health and sustainability in crop production, avoid stress situations, and help in the selection of crops or intercropping systems to improve soil quality.

Chapter 1

Phosphorus-Solubilizing Microorganisms: Perspectives and Challenges

Rodrigo Nogueira de Sousa*
Department of Soil Science, "Luiz de Queiroz" College of Agriculture,
University of São Paulo, Piracicaba, São Paulo, Brazil

Abstract

Phosphorus (P) is the highly sought-after element in tropical soil-based crop systems, primarily attributed to its inherently low bioavailability and limited occurrence in natural deposits. Given the scarcity and finite nature of these deposits, extensive research has been conducted on harnessing the potential of phosphorus solubilizing microorganisms (PSM) to enhance phosphorus availability to plants. These microorganisms solubilize phosphorus present in soils, converting it into plant-assimilable forms. The objective of this chapter is to elucidate the role of PSM in the solubilization of insoluble phosphate and the availability of phosphorus to plants. Furthermore, the chapter delved into the methodologies employed for the selection and isolation of phosphorus-solubilizing microorganisms (PSM). In-depth discussions ensued regarding the diverse applications of PSM in agriculture, with a particular focus on augmenting food production and remediating contaminated soils. Within this context, the identification of microorganisms proficient in solubilizing inorganic phosphorus, along with their associated mechanisms such as organic acid secretion, H+ proton extrusion, and siderophore production, was thoroughly examined. Notably, enzymatic processes were found to be the primary means by which organic phosphorus is mineralized. It became evident that

* Corresponding Author's Email: rodrigoagroufv@gmail.com.

In: Agricultural Soils
Editors: R. Nogueira de Sousa, T. Nascimento Pessoa and K. Ferreira Mendes
ISBN: 979-8-89113-000-5

harnessing the potential of PSM holds immense promise for facilitating the enhanced accessibility of phosphorus to plants, thereby bolstering agricultural efficiency and productivity.

Keywords: microbes, phosphate, organic acids, enzymes, agriculture

Introduction

Phosphorus (P) is a vital nutrient required for the synthesis of essential biomolecules, including nucleic acids, amino acids, lipids, and carbohydrates. However, in most soil environments, phosphorus exists in an insoluble form, predominantly as apatites, which exhibit limited accessibility to plants and other organisms. Fortunately, phosphorus-solubilizing microorganisms (PSM) have emerged as key players in the solubilization of phosphorus. Through their metabolic activities, PSM effectively converts insoluble phosphate into soluble forms, thus enhancing phosphorus uptake by plants (Lim et al., 2007, Kaur et al., 2022). Another prominent instance of microorganisms pertains to Arbuscular Mycorrhizal Fungi (AMF), which possess a remarkable capacity to augment phosphorus assimilation in plants, particularly in nutrient-depleted soils with high leaching tendencies (Sousa, 2023).

One additional advantage associated with the utilization of phosphorus-solubilizing microorganisms (PSM) lies in their capacity to unlock phosphorus that is strongly adsorbed in the soil, originating from diverse compartments such as iron and aluminum oxides, as well as calcium complexes. This microbial action is primarily facilitated through the secretion of organic acids, including but not limited to citric, oxalic, gluconic, and ascorbic acids. These organic substances play a pivotal role in rendering phosphorus available through enzymatic activities, particularly phosphatases, which catalyze the hydrolysis of phosphate esters, thereby liberating soluble phosphate compounds. This intricate mechanism allows for enhanced accessibility of phosphorus, overcoming its previously immobilized state and ensuring its effective utilization by plants.

Microorganisms employ a strategic approach to solubilize phosphorus primarily for their nutritional requirements, expending energy in the process. However, to effectively meet plant demand, these microorganisms ideally possess the ability to solubilize P forms that are typically inaccessible to plants. Figure 1 provides a schematic representation of two distinct pathways of P acquisition by phosphorus-solubilizing microorganisms. In Scenario A, an optimal situation is depicted where microorganisms release excess P, potentially derived from fertilizers, directly to plants. This colonization of soil niches or the rhizosphere by PSM ensures a continuous supply of P to plants in sync with their requirements. Put simply, PSM function as a timely and reliable source of P supplementation for plant nutrition in this scenario. In Scenario B, where there is no excess P, the system is balanced. Microorganisms primarily solubilize P for their utilization, according to their specific needs. Over time, once the microorganism's biomass cycles, P is gradually released into the soil. If microorganisms do not require additional P, they are unlikely to allocate energy toward their solubilization.

Understanding these dynamics of P acquisition and release by microorganisms offers valuable insights into the intricate interplay between microbial activity and plant nutrient availability. By harnessing the potential of PSM, agricultural systems can enhance nutrient cycling and optimize plant uptake for improved productivity and sustainability (Meyer et al., 2019).

The PMS encompasses prokaryotic organisms, including bacteria and fungi, capable of capturing, displacing, or releasing phosphorus in a soluble form under specific environmental conditions. P holds crucial significance as an essential element for plant growth (Karan et al., 2009), and the solubilization performed by PSM plays a pivotal role in making P accessible for uptake by plants and other organisms within the ecosystem. Bacteria have been proposed to play a more dominant and efficient role in this process compared to fungi (Ingle and Padole, 2017). This biological process bridges the gap between P availability and its utilization by living organisms, thus contributing to enhanced nutrient acquisition and overall ecosystem functioning. Three primary categories of PSM can be identified, based on their solubilization mechanisms: those employing organic acids, those utilizing inorganic acids, and those employing enzymes. Organic acids function as solubilizing agents, facilitating the breakdown of insoluble phosphorus (P) compounds, while inorganic acids, including phosphoric acid, play a role in converting insoluble P into a soluble form. Additionally, PSM harbors phosphorus-solubilizing enzymes, such as acid and alkaline phosphatase, as well as phospholipase, which contribute to the solubilization process (Silva et al., 2008). These diverse mechanisms employed by PSM highlight their versatile capacity to enhance P availability in various environmental contexts, supporting nutrient cycling and plant productivity.

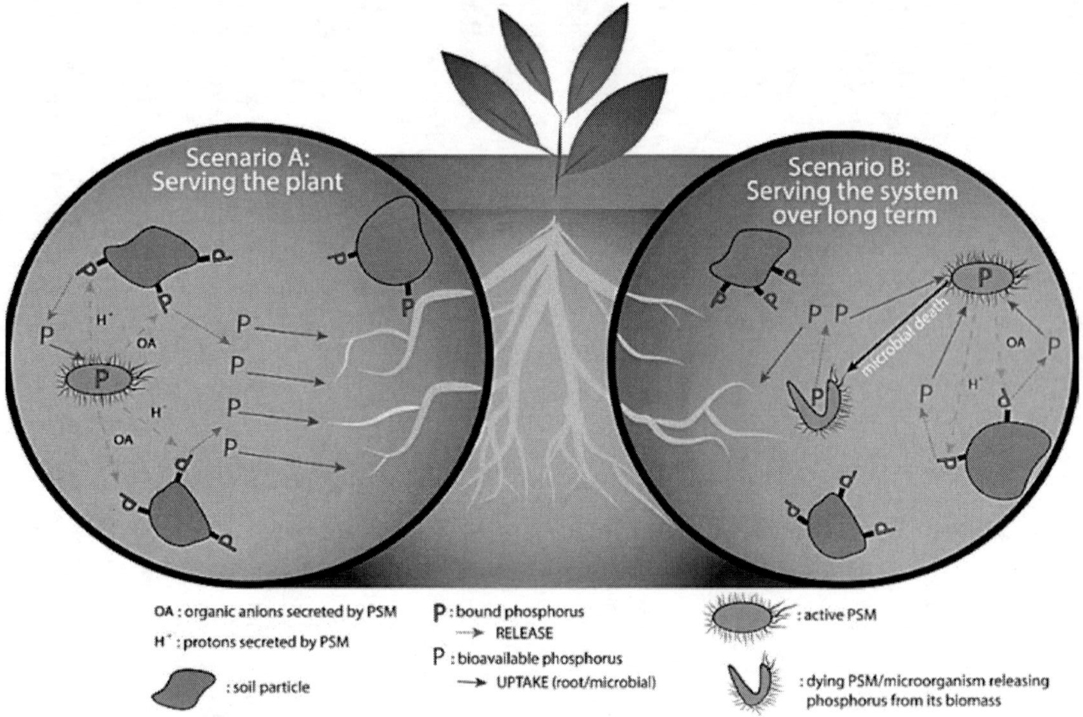

Source: Adapted from Raymond et al., (2021).

Figure 1. Summary of two theoretical processes by which phosphate-solubilizing microorganisms' function (PSM). The ideal PSM contribution to plant P nutrition is represented by Scenario A. Scenario B shows how PSM release P from their biomass turnover to maintain their individual needs and satisfy the process over the long term.

In agricultural production systems, PSMs are used to increase the amount of phosphorus that is available for plant uptake. They are also used in the remediation of contaminated soils since they can solubilize and release P from insoluble compounds, making this nutrient available for uptake by other organisms (Park et al., 2011). PSMs are also used in water treatment systems to assist in the removal of P from bodies of water, making them safer to drink (Kim et al., 2005). As we can see, these PSMs are beneficial organisms for the environment and agriculture.

They are used to improve soil health, water quality, and food production. As these organisms are easily found in nature, they are easy to manipulate as a tool for improving soil fertility and water quality.

The primary aim of this chapter is to provide a comprehensive understanding of the crucial role played by PSM in the solubilization of insoluble phosphate compounds and the subsequent availability of P to plants. Furthermore, the chapter delves into the methods employed for the selection and isolation of PSM, along with the various techniques utilized to assess their solubilizing capacity. Lastly, the chapter explores the diverse applications of PSM in agriculture, with a particular focus on enhancing food production and facilitating the remediation of contaminated soils. By shedding light on these aspects, this chapter aims to underscore the immense potential of PSM in sustainable agriculture and soil management practices.

Organic and Inorganic P Fractions in Soil

In the natural environment, phosphorus (P) exists in both organic (Po) and inorganic (Pi) forms. The organic form originates from plant and animal waste, encompassing macro- and microfauna contributions, while the inorganic form primarily derives from phosphate rock, manifesting as phosphate ions in solution. Phosphate ions in solution can exist in the concentrated states of H_3PO_4, $H_2PO_4^-$, HPO_4^{-2}, and PO_4^{-3}, with their respective concentrations varying according to the soil pH. According to Harrison (1987), the organic form accounts for approximately 30% to 65% of the total P, while the inorganic form comprises 35% to 70%. Tisdale et al., (1993), Nahas et al., (1994), and Rodríguez and Fraga (2002) estimate that Po constitutes 10–50% of total soil P. Within the organic form, inositol-phosphates, make up 10% to 50%, phospholipids comprise 1% to 5%, and nucleic acids constitute 0.2% to 0.5%. In contrast, Pi is associated with insoluble mineral phosphates, phosphate anions, iron and aluminum hydroxides, aluminum silicates, and calcium carbonates. Conversely, Po is closely linked to microorganisms and plants, which release enzymes capable of breaking down formed particles. As a defense mechanism, Po forms complexes between isositol-phosphates and proteins to withstand these adversities.

In Brazilian Cerrado soils, for example, Pi concentrations in solution are naturally below 0.1 ppm (Goedert et al., 1986). This value is negligible (insignificant) due to some factors, such as the characteristic mineralogy of soils, i.e., many soils come from originally poor parent material, are very weathered soils, have low clay contents, and have high leaching of cations. All this chemical "poverty" causes the soils to present a high fixation of P ions in iron and aluminum oxides, which end up presenting a low concentration in the solution.

The classification of Po and Pi can be subdivided into labile and non-labile P (Guerra et al., 1996). The labile fraction is represented by the set of compounds capable of quickly replacing the soil solution and being absorbed by plants or microorganisms, while the non-labile fraction is the P strongly adsorbed to the soil colloids, with high binding energy, or that are part of the mineral structure and become slow to be released due to the low solubility of the compounds. As mentioned, all this variation in availability will depend on the degree of soil weathering, the mineralogy, organic matter content, physicochemical characteristics, biological activity, and the predominant vegetation.

P Solubilizing Activity by Microorganisms

As previously discussed, the solubilization of phosphorus (P) is a crucial process for plant nutrition. This vital function is primarily executed by microorganisms, including bacteria and fungi, which secrete enzymes such as acid phosphatase. These enzymes play a pivotal role in disrupting the stable bonds between phosphorus and other minerals, thereby rendering them accessible for plant uptake (Amorim et al., 2017).

The investigation of P solubilizing activity by microorganisms encompasses various crucial aspects. These include the identification of the specific microorganisms involved in this process, the characterization of the enzymes they produce, the assessment of enzyme activity under diverse conditions, and the analysis of the impact of abiotic factors (such as temperature, pH, humidity, etc.) on the solubilization process. Furthermore, this research may delve into the production of organic compounds that facilitate P solubilization, as well as explore the potential application of these microorganisms in fertilization programs. By exploring these facets, a comprehensive understanding of P solubilization can be attained (Silva et al., 2017).

As already mentioned, P solubilization is the process by which microorganisms release existing P to plants in non-usable forms. Microorganisms employ several key mechanisms to enhance soil phosphorus solubility, including the secretion of mineral-solubilizing compounds, enzymatic-mediated biochemical phosphate mineralization, and phosphorus assimilation through the release of phosphorus from complex substrates. This process is important for plant nutrition and growth since the P present in the soil is generally poorly soluble. PSM has been studied in several countries around the world, including Brazil, USA, Canada, India, and Australia (Vieira et al., 2014). Therefore, research must be continuously conducted to improve techniques for PSM around the world.

Several organic acids are produced by PSM (Table 1). 2-Cetogluconic acid was identified in strains of *Rhizobium leguminosaru* bv viceae that showed solubilization capacities between 47 and 52% of phosphate rock in yeast glucose culture medium after 3 days of incubation (Halder et al., 1990).

Table 1. Organic acids produced by phosphate-solubilizing microorganisms (PSM)

Phosphate-solubilising microorganism (PSM)	Metabolites (organic acids)	Reference
Penicillium rugulosum	Citric and gluconic	Reyes et al., (2001)
Acinetobacter sp.	Oxalic, gluconic, and butyric	Puente et al., (2009)
Enterobacter intermedium	Gluconic	Hwangbo et., (2003)
Pseudomonas fluorescens	Citric, malic, tartaric, and gluconic	Fankem et al., (2006)
Paenebaccilus kribbensis	Gluconic, oxalic, and propionic	Marra et al., (2012)
Aspergillus niger and Penicillium	Citric, oxalic, and gluconic	Mendes et al., (2013)
Psedomonas nitroreducens	Indole aceti	Pemila et al., (2014)
Aspergillus japonicus	Salicylic, ascorbic, citric, formic, lactic, oxalic, and malic	Nasr et al., (2021)

Isolation of New Species of PSM

P is an essential element for life and is applied to the soil for plants as fertilizer. However, the vast majority of the P in the soil is in inorganic form and strongly adsorbed to the colloid, making it unavailable to plants. However, as seen, PSM can reverse this effect to a certain extent, making it available to plants.

To isolate new species of PSM, the following steps can be followed (Ribeiro et al., 2010):

1. Collection of soil samples. For this, soil samples should be collected from different types of environments, such as forests, pastures, agricultural fields, etc.
2. Sample preparation. The samples should be treated properly to remove any solid particles present in the soil.
3. Cultivation of microorganisms. Samples should be grown in suitable culture media for the proliferation of microorganisms.
4. P solubilization tests. Cultured microorganisms should be subjected to tests to determine their ability to solubilize inorganic P.
5. Isolation of new species. The species of microorganisms that show a high capacity for P solubilization should be isolated and identified.

The isolation of new species of PSM is important to understand how these organisms affect P solubilization and to develop new fertilization strategies. Isolating these species involves collecting samples from different sources, such as soil, water, and other environments, and performing microorganism cultivation techniques (Ribeiro et al., 2010).

Mohamed et al., (2019), in their study, used the serial dilution method and growing on selective media (National Botanical Research Institute's phosphate growth medium - NBRIP), bacteria from various soil samples for phosphate solubilizing ability. In the presumptive test, bacterial colonies with distinct halo zones surrounding the microbial growth were taken into consideration as phosphate solubilizers (Figure 2). The clear zone surrounding the colonies was re-screened to determine the amount of P that was present in the medium.

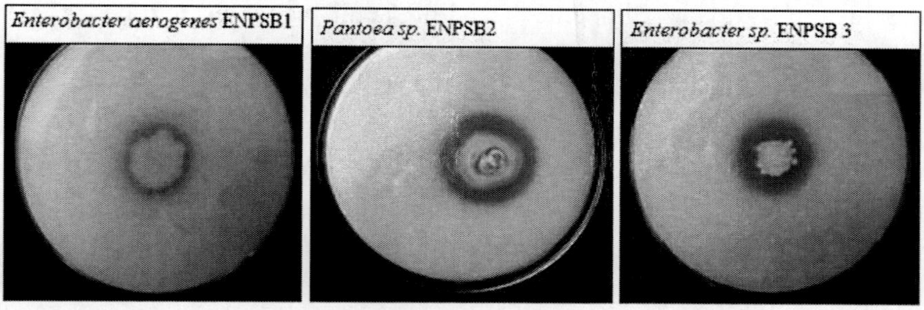

Source: Adapted from Mohamed et al., (2019).

Figure 2. Bacterial colonies showing clear halo zones around the microbial growth of novel phosphate dissolving bacterial isolates.

After isolation, the species of PSM should be characterized through physiological, biochemical, and molecular analyses to determine their characteristics and ability to solubilize P. Some important characteristics include the ability to produce organic acids, the identification of enzymes related to P solubilization, the identification of genes associated with the process of P solubilization, and the ability to make use of P sources (Ribeiro et al., 2010).

Nutritional and Biochemical Action of PSM

The nutritional and biochemical activities of PSM are influenced by various factors, including the specific microorganism involved, soil characteristics, nutrient availability, and the presence of other microorganisms in the soil. To gain insights into the nutritional profile of PSM, it is essential to investigate their specific requirements for macronutrients such as carbon, nitrogen, and phosphorus, as well as micronutrients like iron and zinc, among others (Bhargava et al., 2012). This comprehensive characterization enables a deeper understanding of the nutrient dynamics and metabolic activities of PSM, enhancing our knowledge of their role in soil fertility and plant nutrition.

The activity of PSM plays a pivotal role in promoting plant growth, particularly in soils characterized by limited phosphorus availability. Beyond their capability to solubilize phosphorus, these microorganisms also exhibit the remarkable ability to fix atmospheric nitrogen, thereby enhancing soil fertility and fostering favorable microbiological activity (Charana 2012). Consequently, harnessing the potential of PSM holds significant promise for sustainable agricultural practices and the cultivation of nutrient-efficient crops.

The biochemical characterization of PSM involves the identification of various enzymes and other molecules related to P solubilization, as well as the characterization of their synthesis and regulation mechanisms. Another important aspect of the nutritional and biochemical characterization of PSM concerns the evaluation of their capacity for growth and enzyme production. Generally, this evaluation is carried out in the laboratory, using solid or liquid cultures.

PSM are frequently classified as bacteria, fungi, or algae (Sharma et al., 2013). P-solubilizing bacteria, for example, are generally classified as Gram-negative bacteria and may include Bacillus, Pseudomonas, Enterobacter, and Acinetobacter. These bacteria produce enzymes in soluble forms, phosphoric acid, and phosphoglyceric acid, and use these compounds for energy.

P-solubilizing fungi are also Gram-negative and can include species such as Penicillium, Aspergillus, Trichoderma, and Glomus (Kalayu 2019). These fungi also produce enzymes that solubilize P, and the nutrients are used to produce energy. In addition, P-solubilizing fungi can also produce hormones that promote plant growth.

Several studies have been conducted to improve P solubilization techniques using microorganisms. For example, some studies have shown that the use of *Rhizobium* sp. bacteria, *Pseudomonas fluorescens*, and *Bacillus* sp. were effective in P-solubilizing. In addition, it has also been shown that the addition of organic compounds, such as citric acid, can increase P solubilization (Silva et al., 2018). It is known that the addition of *Bacillus subtilis* bacteria to soil can increase P solubilization on average by up to 4-fold, as well as that the addition of *Bacillus megaterium* bacteria can increase phosphorus solubilization by up to 2-fold and the addition of mycorrhizal fungi can increase phosphorus solubilization by up to 3-fold. Furthermore, PSM is more prominent in soils with low levels of organic matter. This behavior can be attributed to the scarcity of nutrients, reduced microbial competition, and the ability of PSM to actively solubilize phosphorus, offering a significant advantage for plant growth in nutrient-limited environments.

P-solubilizing algae are generally classified as cyanobacterial species, such as Anabaena, Oscillatoria, Microcystis, and Spirulina (Zahra et al., 2020). These algae also produce enzymes capable of P-solubilizing and use the solubilized nutrients for energy. PSMs are important for maintaining soil fertility and can be used to increase crop productivity in general. In addition, these microorganisms are also important for water purification and wastewater treatment.

Study of Environmental Variables in the PSM

PSMs play a crucial role in global nutrient cycling, contributing to the availability and cycling of phosphorus in various ecosystems (Prasad et al., 2021). The activity of PSM is greatly influenced by environmental variability, which encompasses a range of factors that impact their function. Environmental variables, such as temperature, pH, conductivity, nutrient levels, salinity, and others, are assessed in studies investigating the influence of these factors on PSM dynamics. Through the examination of these variables, a comprehensive understanding of the environmental conditions that shape PSM activity and phosphorus availability can be obtained, aiding in the development of sustainable agricultural practices and nutrient management strategies.

The study involves the sampling and analysis of soil and water at different locations. PSMs are identified and isolated from the samples. The effects of different environmental variables are evaluated through laboratory experiments, including assessing the effect of different variables on PSM. The results provide important information on how environmental variables influence the PSM and important information for the proper management of water resources and soils.

Study of the Influence of PSM on Plant Growth

P is a vital nutrient for plant growth; however, a significant portion of soil P is often inaccessible to plants, particularly in tropical regions. This is where phosphorus-solubilizing microorganisms play a crucial role by producing enzymes and organic acids.

The investigation into the influence of these microorganisms on plant growth encompasses various aspects, including the identification and isolation of these microorganisms, characterization of their properties, assessment of their effectiveness in solubilizing soil P, and evaluation of their impact on plant growth (Rodrigues and Fraga, 1999).

In order to assess the influence of PSM on plant growth, comprehensive experiments can be conducted. For instance, plants can be cultivated in soils with varying levels of PSM diversity and abundance. The outcomes of these experiments can be evaluated through growth tests, encompassing parameters such as plant height, biomass, chlorophyll index, and other relevant indicators. Furthermore, the levels of phosphorus in the soil can be analyzed to determine if there has been successful nutrient solubilization. By examining the results of such studies, we can gain a deeper understanding of the impact of PSM on plant growth. This valuable information holds immense significance for sustainable agricultural practices, as it enables the utilization of PSM to enhance nutrient availability in the soil. With a comprehensive understanding of the interplay between PSM and plant growth, we can pave the way for more effective strategies in agricultural production, fostering nutrient optimization and contributing to the advancement of sustainable food systems.

Application of PSM in Agricultural Systems

As mentioned throughout this chapter, it has been seen that bacteria and fungi are part of the process of solubilization of P in the soil (Paul and Clark 1996, Richardson et al., 2001). It can be said that these microorganisms are part of the biogeochemical cycle of P and are therefore responsible for the transfer of P between soil attachment sites. The tendency to solubilize phosphate, inorganic or organic, is a characteristic of many of the bacteria present in the soil in association with the rhizosphere of various plant species.

These microorganisms perform essential and effective metabolic processes for the solubilization and mineralization of phosphate from poorly available forms of inorganic P (phosphate rocks) and organic P (organic matter). Among the various microorganisms that perform these processes, bacteria have the greatest potential for obtaining soluble phosphate. It is estimated that about 40% of rhizospheric bacteria can solubilize P (Richardson, 2001). According to Mendes (2003) these microorganisms directly impact the ability of plants to acquire available P in the soil through various mechanisms, among them:

- Increase of root surface area by the extension of the root system (mycorrhizal associations);

- Promotion of lateral root growth and/or root hairs (through phytohormones);
- Adsorption equilibrium shift (allows for the transfer of phosphate ions or organic phosphate).

Through solubilization and mineralization mechanisms, PSMs can convert inorganic and organic P into soluble forms (Khan et al., 2009). This affects the bioavailability of P and, as a consequence, facilitates its uptake by plant roots. The use of the solubilization mechanism is necessary for optimal performance within varying field conditions. Thus, understanding the P cycle in the soil as well as in plants and microorganisms is indispensable.

PSMs also assist in improving soil structure by increasing porosity, which increases water storage in the soil, which is beneficial for plant growth (Helliwell et al., 2014). In addition, the use of these microorganisms helps reduce the amount of phosphate fertilizer applied, saving costs and reducing dependence on imports, as well as assisting in reducing environmental pollution.

The application of these microorganisms is usually done using commercial inoculants. These inoculants contain the types of solubilizing microorganisms, such as P-solubilizing bacteria, that are added to the soil. This improves the availability of P in the soil, providing nutrients for plants and helping to improve crop production. The addition of PSM also helps to improve root growth, increase drought resistance, and helps reduce soil erosion.

In some farming systems, such as those where the soil is deficient in phosphorus, farmers may also use phosphate fertilizers to improve phosphorus availability. However, indiscriminate application of phosphate fertilizers can lead to P accumulation in the soil, which can have a negative impact on soil quality and the environment, particularly on sandier soils with low organic matter content. In addition, higher fertilizer application rates result in higher production costs. Therefore, the use of PSM is a good alternative to maximize the availability of P present in the soil for the solution.

Study of the Interaction PSM Fungi

The fungus *Aspergillus niger* is one of the most studied given its tendency to produce organic acids, especially citric acid (Bizukojc and Ledakowicz 2004).

Some studies use the production of oxalic acid by *A. niger*, which makes it a very active agent for iron and aluminum phosphate solubilization, and gluconic acid, for calcium phosphate solubilization (Chuang et al., 2007). It is also important to emphasize the importance of arbuscular mycorrhizal fungi (AMF), which act as extensions of the root system, causing an increase in the volume of soil explored and the speed of P uptake. These microorganisms can release P from minerals and form soluble organic compounds that can be absorbed by plants. The solubilized P can also be taken up by the fungi, which act as transport agents between the nutrient sources in the soil and the plants.

The investigation of PSMs holds great significance as it enhances our understanding of soil nutrient cycling and contributes to the improvement of soil fertility, a crucial factor for agricultural production. The addition of PSMs to the soil has been shown to augment the availability of nutrients for plants, thereby stimulating their growth. Moreover, this symbiotic relationship also benefits fungi, as they receive a vital supply of phosphorus from these microorganisms, essential for their growth. By exploring the intricate interactions between fungi and other PSMs, we gain valuable insights into nutrient dynamics within the soil ecosystem and pave the way for enhanced fertility management strategies. This knowledge is paramount for sustainable agriculture.

Use of PSM As Biofertilizers

The use of inoculants, based on the use of microorganisms, according to Rajapaksha et al., (2011), has been promising in the development of crops such as rice, reducing the use of phosphate fertilizers by up to 50%. Oliveira et al., (2009) point out several advantages of the use of this type of product, such as the availability of P complexed with aluminum, iron, and calcium and the dissolution of phosphate rock materials present in the medium, besides the release of organic acids. Mander et al., (2012) and Greiner (2007) complement this idea by adding the occurrence of the release of phosphatases and phytases, which promote the mineralization of organic P present in the soil.

The use of PSM as a biofertilizer is extremely important for the healthy development of crops. P is an essential element for plant nutrition, as it is responsible for processes such as photosynthesis, protein formation, respiration, growth, and reproduction. However, most of the P present in highly weathered soils is known to be insoluble or strongly adsorbed in minerals such as kaolinite, hematite, gibbsite, and others (Haynes 1982). This hotspot is where PSM becomes useful. In addition, microorganisms also improve soil structure, provide other nutrients, and reduce the toxicity of heavy metals present in the soil. As already mentioned, it is first necessary to identify and select the appropriate microorganisms for the process.

Once a suitable type of PSM is selected, it can be inoculated into the soil. The inert inorganic P is then solubilized and released to the plant. These microorganisms can also be applied as foliar fertilizer in the form of a liquid culture medium. The use of PSMs as biofertilizers has proven to be an effective alternative to crop fertilization because they are readily available and can be safely applied without the need for inputs.

It is important to note that while PSM can increase the amount of phosphorus available to the plant, it does not increase the amount of other important nutrients such as nitrogen (except for rhizobia), calcium, magnesium, and potassium. Therefore, the addition of commercial fertilizers containing these nutrients is recommended.

Study of the Diversity of PSM

P is one of the most important nutrients for living things, which is essential to produce energy, cell structures, and other important molecules. As such, P solubilization is an important process for the availability of this nutrient.

The study of the diversity of solubilizing microorganisms aims to identify the species of microorganisms that can solubilize P. These species are usually identified through molecular biology techniques, such as DNA and RNA sequence analysis (Igual et al., 2001). These techniques allow the identification of organisms through the analysis of their genomes, as well as the comparison between different species. In addition, studies of microorganism genetics can be performed to evaluate the potential of certain species to solubilize P.

This approach allows for the identification of potential PSMs and the understanding of the diversity of these organisms and their characteristics. Furthermore, this research can provide information on the environmental conditions required for P solubilization and the influence of these microorganisms on soil nutrient cycling. These studies are important for the development of new techniques to improve soil fertility and increase the efficiency of phosphate fertilizer use.

Study of Abiotic Stress Tolerance of PSM

Abiotic stress tolerance of PSM is the ability of these microorganisms to withstand adverse environmental conditions, such as temperature, pH, salinity, and ultraviolet radiation, and continue to solubilize P from insoluble sources (Khan et al., 2021). This tolerance is important to allow microorganisms to be used in different environments and conditions. Abiotic stress is an adaptive response to external factors capable of altering an organism's environmental conditions. PSM are organisms that produce soluble molecules capable of releasing P from insoluble forms present in the soil.

Studying the abiotic stress tolerance of PSM can help us understand how these organisms adapt to the various environmental factors that can affect their activity. This research could examine how PSM responds to changes in temperature, salinity, pH, UV radiation, and other variables. This information can be used to develop strategies to improve the performance of these organisms under stressful conditions. In addition, studying the abiotic stress tolerance of PSM can help develop strategies for the conservation and management of these microorganisms in different environments.

Importance and Perspectives of PSM

The importance of PSM for agricultural production systems is widely recognized and has been highlighted over the years. P is essential for plant growth, as it acts as an indispensable nutrient for cellular metabolism. In addition, phosphorus also plays an important role in energy production, protein synthesis, and the regulation of important processes such as respiration.

The adequate presence of PSM in the soil is known to be essential for improving plant growth and crop yields, as it enables plants to obtain the nutrients they need efficiently. Furthermore, PSM improves soil quality by preventing soil erosion and increasing organic matter and nutrient availability. In addition, the presence of these microorganisms also helps to reduce food production costs and external dependence on phosphate fertilizer imports.

Conclusion

Phosphorus-solubilizing microorganisms (PSM) are an important tool for increasing nutrient availability to plants. They help make the P in solution available to plants, which use it to produce energy and other essential compounds. These microorganisms can also help reduce the negative effects of P leaching, which can contribute to the eutrophication of surface and groundwater.

However, it is important to remember that PSMs are only part of the solution. The use of proper agronomic techniques, such as the use of adequate and correct fertilizers, the addition of organic matter, no-till farming, cover crops, crop rotation, proper irrigation water management, and erosion control, are all key to maintaining P and other nutrient reserves in the soil. The use of PSM can be an important tool for the efficient management of water resources in arid or semi-arid conditions.

PSMs are dependent on environmental factors such as temperature, humidity, pH, and the nutritional balance of other nutrients. Therefore, for the best results, proper management is required to ensure that these microorganisms are effective. In summary, PSM should only be used as part of a more comprehensive nutrient management program that includes proper agronomic techniques and the use of appropriate fertilizers.

Acknowledgments

The author thanks the Foundation for Research Support of the State of São Paulo (FAPESP - 2020/13710-0). The author also thanks the National Council for Scientific and Technological Development (CNPq - 168192/2018-7).

References

Amorim, K. M. S., Pereira, J. S., Andrade, S. C. 2017. Estudo da atividade solubilizante de fósforo por microrganismos isolados do cerrado [Study of phosphorus solubilizing activity by microorganisms isolated from the Cerrado]. *Revista Brasileira de Ciências Agrárias* [*Brazilian Journal of Agricultural Sciences*] 12(2)124–131.

Bhargava, A., Carmona, F. F., Bhargava, M., Srivastava, S. 2012. Approaches for enhanced phytoextraction of heavy metals. *Journal of Environmental Management* 105, 103–120.

Bizukojc, M., Ledakowicz, S. 2004. The kinetics of simultaneous glucose and fructose uptake and product formation by *Aspergillus niger* in citric acid fermentation. *Process Biochemistry* 39(12)2261-2268.

Carvalho, S. T., Vieira, S. F., Pires, M. R., Silva, J. P. M. 2011. Solubilização de fósforo em solo por bactérias endofíticas associadas a raízes de cana-de-açúcar. [Solubilization of phosphorus in soil by endophytic bacteria associated with sugarcane roots] *Revista Brasileira de Ciência do Solo [Brazilian Journal of Soil Science]* 35(4)1279-1292.

Charana, W. B. 2012. Prospectus of phosphate solubilizing microorganisms and phosphorus availability in agricultural soils: A review. *African Journal of Microbiology Research* 6(37)6600-6605.

Chuang, C. C., Kuo Y. L., Chao, C. C., Chao, W. L. 2007. Solubilization of inorganic phosphates and plant growth promotion by *Aspergillus niger*. *Biology and Fertility of Soils* 43(5)575-584.

Fankem, H., Nwaga, D., Deube, A., Dien, L., Merbach, W., Etoa, F. X. 2006. Occurence and functioning of phosphate solubilizing microorganisms from oil palm tree (*Elaeis guineensis*) rhizosphere in Cameroon. *African Journal of Biotechnology* 5, 2450-2460.

Goedert, W., Sousa, D. M., Lobato, E. 1986. Fósforo. In: Goedert, W. J. *Solos do Cerrados: tecnologias e estratégias de manejo.* [Cerrado soils: technologies and management strategies.] Planaltina, DF, Brazil: Embrapa- CPAC; São Paulo, SP, Brazil: Nobel, p. 129-186.

Greiner, R. 2007. Phytate-degrading enzymes: regulation of synthesis in microorganisms and plants. In: Turner, B. L.; Richardson; A. E.; Mullaney, E. J. *Inositol phosphates: linking agriculture and the environment.* Wallingford, Oxfordshire, UK: Cabi, p. 78-96.

Guerra, J. G. M., Almeida, D. J., Santos, G. A., Fernandes, M. S. 1996. Conteúdo de fósforo orgânico em amostras de solo. [Organic phosphorus content in soil samples.] *Pesquisa Agropecuária Brasileira* [*Brazilian Agricultural Research*] 31(4)291-299.

Halder, A. K., Mishra, A. K., Bhattacharyya, P., Chakrabartty, P. K. 1990. Solubilization of rock phosphate by Rhizobium and Bradyrhizobium. *The Journal of General and Applied Microbiology* 36(2)81-92.

Haynes, R. J. 1982. Effects of liming on phosphate availability in acid soils. *Plant and Soil* 68(3)289–308.

Helliwell, J., Miller, A., Whalley, W., Mooney, S., Sturrock, C. 2014. Quantifying the impact of microbes on soil structural development and behaviour in wet soils. *Soil Biology and Biochemistry* 74, 138–147.

Hwangbo, H., Park, R. D., Kim, Y. S., Park, K. H., Kim, T. H., Suh, J. S., Kim, K. Y. 2003. 2-ketoguconic acid production and phosphate solubilization by Enterobacter intermedium. *Current Microbiology* 47(2)87-89.

Ingle, K. P., & Padole, D. A. 2017. Phosphate solubilizing microbes: An overview. *International Journal of Current Microbiology and Applied Sciences*, 6(1), 844-852.

Igual, J. M., Valverde, A., Cervantes, E., Velázquez, E. 2001. Phosphate-solubilizing bacteria as inoculants for agriculture: use of updated molecular techniques in their study. *Agronomie* 21(6–7)561–568.

Kalayu, G. 2019. Phosphate solubilizing microorganisms: promising approach as biofertilizers. *International Journal of Agronomy*, 1–7.

Karan, S., Singh, J., Singh, B. 2009. Potencial de solubilização de fósforo de microrganismos do solo. [Phosphorus solubilization potential of soil microorganisms.] *Revista Brasileira de Ciência do Solo [Brazilian Journal of Soil Science]* 33(1)81-90.

Kaur, R., Kaur, S. 2022. Exploration of phytate-mineralizing bacteria with multifarious plant growth-promoting traits. *BioTechnologia* [*Biotechnology*] 103(2)99–112.

Khan, A. A., Jilani, G., Akhtar, M. S., Naqvi, S. M. S., Rasheed, M. 2009. Phosphorus solubilizing bacteria: occurrence, mechanisms and their role in crop production. *Journal of Agriculture and Biological Sciences* 1(1)48-58.

Khan, N., Ali, S., Shahid, M. A., Mustafa, A., Sayyed, R. Z., Curá, J. A. 2021. Insights into the Interactions among roots, rhizosphere, and rhizobacteria for improving plant growth and tolerance to abiotic stresses: a review. *Cells* 10(6)1551.

Kim, Y. H., Bae, B., Choung, Y. K. 2005. Optimization of biological phosphorus removal from contaminated sediments with phosphate-solubilizing microorganisms. *Journal of Bioscience and Bioengineering* 99(1)23–29.

Lim, B. L., Yeung, P., Cheng, C., Hill, J. E. 2007. Distribution and diversity of phytate-mineralizing bacteria. *The ISME Journal* 1(4)321–330.

Mander, C., Wakelin, S., Young, S., Condron, L., O'Callaghan, M. 2012. Incidence and diversity of phosphate-solubilising bacteria are linked to phosphorus status in grassland soils. *Soil Biology and Biochemistry* 44(1)93-101.

Marra, L. M., Soares, C. R. F. S., Oliveira, S. M., Ferreira, P. A. A., Soares, B. L., Carvalho, R. F., Lima, J. M., Moreira, F. M. S. 2012. Biological nitrogen fixation and phosphate solubilization by bacteria isolated from tropical soils *Plant and Soil* 20, 1-19.

Mendes, I. C., Reis Junior, F. B. 2003. *Micro-organismos e disponibilidade de fósforo (P) nos solos: uma análise crítica.* [*Microorganisms and phosphorus (P) availability in soils: a critical analysis.*] Embrapa Cerrados-Documentos 85. 24 p.

Mendes. G. O., Dias, C. S., Silva, I. R., Junior, J. I. R., Pereira, O. L., Costa, M. 2013. Fungal rock solubilization using sugarcane bagasse. *World Journal of Microbiology and Biotechnology* 29, 43-50.

Meyer. G., Bünemann, E. K., Frossard, E., Maurhofer, M., Mäder, P and Oberson, A. 2017. Gross phosphorus fluxes in a calcareous soil inoculated with Pseudomonas protegens CHA0 revealed by 33P isotopic dilution. *Soil Biology and Biochemistry* 104: 81– 94.

Mohamed, A. E., Nessim, M. G., Abou-el-seoud, I. I., Darwish, K. M., Shamseldin, A. 2019. Isolation and selection of highly effective phosphate solubilizing bacterial strains to promote wheat growth in *Egyptian calcareous* soils. *Bulletin of the National Research Centre* 43(1)203.

Nahas, E., Centurion, J. F., Assis, L. C. 1994 Micro-organismos solubilizadores de fosfato e produtores de fosfatases de vários solos. [Phosphate solubilizing and phosphatase-producing microorganisms from various soils.] *Revista Brasileira de Ciência do Solo [Brazilian Journal of Soil Science]* 18, 43-48.

Nasr, S., Mousa, A., Yasser, M. 2021. Quantitative and qualitative analysis of organic acids produced by phosphate solubilizing fungi. *Egyptian Journal of Botany* 61(1)167-176.

Oliveira, C. A., Alvez, V. M. C., Marriel, I. E., Gomes, E. A., Scotti, M. R., Carneiro, N. P., Gimarães, C. T., Schaffert, R. E., Sá, N. M. H. 2009. Phosphate solubilizing microorganisms isolated from rhizosphere of maize cultivated in an Oxisol of the Brazilian Cerrado Biome. *Soil Biology and Biochemistry* 41(9)1782-1787.

Park, J. H., Bolan, N., Megharaj, M., Naidu, R. 2011. Isolation of phosphate solubilizing bacteria and their potential for lead immobilization in soil. *Journal of Hazardous Materials* 185(2–3)829–836.

Paul, E. A., Clark, F. E. 1996. *Soil microbiology and biochemistry*. San Diego, USA: Academic Press, 340 p.

Pemila, E., Chitraselvi, R., Kalidass, S., Rajiv, K. 2014. The efficiency of rhizosphere bacteria in the production of indole acetic acid, siderophore, and phosphate solubilization. *International Journal of ChemTech Research* 6, 2557-2564.

Prasad, S., Malav, L. C., Choudhary, J., Kannojiya, S., Kundu, M., Kumar, S., Yadav, A. N. 2021. Soil microbiomes for healthy nutrient recycling. In: Yadav, A. N., Singh, J., Singh, C., Yadav, N. (eds). *Current Trends in Microbial Biotechnology for Sustainable Agriculture.* Singapore: Springer, p. 1-22.

Puente, M. E., Li, C. Y., Bashan, Y. 2009. Rock-degrading endophytic bacteria in cacti. *Environmental and Experimental Botany* 66, 389-401.

Rajapaksha, R. M. C. P., Herath, D., Senanayake, A. P., Senevirathne, M. G. T. L. 2011. Mobilization of rock phosphate phosphorus through bacterial inoculants to enhance growth and yield of wetland rice. *Communications in Soil Science and Plant Analysis* 42(3)301-314.

Raymond, N. S., Gómez-Muñoz, B., van der Bom, F. J. T., Nybroe, O., Jensen, L. R., Müller-Stöver, D. S., Oberson, A., Richardson, A. E. 2021. Phosphate-solubilising microorganisms for improved crop productivity: a critical assessment. *New Phytologist* 229(3)1268-1277.

Reyes, I., Baziramakenga, R., Bemeir, L., Antoun, H. 2001. Solubilization of phosphate rocks and minerals by a wild-type strain and two UV-induced mutants of *Penicillium regulosum. Soil Biology and Biochemistry* 33:1741-1447.

Ribeiro, A. P., Buckeridge, M. S. 2010. Isolamento de bactérias solubilizadoras de fosfato de um solo da região sul do Brasil. [Isolation of phosphate solubilizing bacteria from a soil in southern Brazil.] *Ciência Rural* [*Rural Science*] 40(7)1644-1650.

Richardson, A. E. 2001. Prospects for using soil microorganisms to improve the acquisition of phosphorus by plants. *Functional Plant Biology* 28(9)897-906.

Rodríguez, H., Fraga, R. 1999. Phosphate solubilizing bacteria and their role in plant growth promotion. *Biotechnology Advances* 17(4–5)319–339.

Rodríguez, H., Fraga, R., Gonzalez, T., Bashan, Y. 2002. Genetic modifications of phosphate solubilization bacteria to be used as agricultural inocculants. In: *International Meeting on Microbial Phosphate Solubilization.* Salamanca, Spain.

Sharma, S. B., Sayyed, R. Z., Trivedi, M. H., Gobi, T. A. 2013. *Phosphate solubilizing microbes: sustainable approach for managing phosphorus deficiency in agricultural soils.* SpringerPlus 2(1)587.

Silva, C. M., Barros, N. F., Jorge, J. A., Alvarenga, A. A., Araújo, W. L., Guimarães, L. T. S., Melo, I. S. 2008. Atividade solubilizadora de fósforo de bactérias do solo de área de pastagem. [Phosphorus solubilizing activity of soil bacteria in pasture area.] *Revista Brasileira de Ciência do Solo [Brazilian Journal of Soil Science]* 32(2)657-666.

Silva, D. A., Costa, L. C., Dias, D. C. 2018. Estudo da atividade solubilizante de fósforo por microrganismos do solo. [Study of phosphorus solubilizing activity by soil microorganisms.] *Revista Brasileira de Ciências Agrárias* [*Brazilian Journal of Agricultural Sciences*] 13(1)75–81.

Silva, J. P., Fonseca, E. V., Dias, D. C. 2017. Estudo da atividade solubilizante de fósforo por bactérias isoladas de solos de Cerrado. [Study of phosphorus solubilizing activity by bacteria isolated from Cerrado soils] *Revista Brasileira de Ciências Agrárias* [*Brazilian Journal of Agricultural Sciences*] 12(3)176–184.

Sousa, R. 2023. Introductory Chapter: Mycorrhizal Fungi – A Current Overview on Agricultural Productivity and Soil Health [Printed]. In: Sousa, Rodrigo (Ed). *Arbuscular Mycorrhizal Fungi in Agriculture - New Insights*. United Kingdom: Intechopen. https://doi.org/10.5772/intechopen.109021.

Tisdale, S., Nelson, W. E., Beatan, J. D., Havlin, J. L. 1993. *Soil fertility and fertilizers.* 5th ed, New York, USA: Mac Millan. 648 p.

Vieira, A. F., Carvalho M. B., Silva, V. A. 2014. Estudo da atividade solubilizante de fósforo por fungos filamentosos. [Study of phosphorus solubilizing activity by filamentous fungi] *Revista Brasileira de Ciências Agrárias* [*Brazilian Journal of Agricultural Sciences*] 9(1)88–95.

Zahra, Z., Choo, D. H., Lee, H., Parveen, A. 2020. Cyanobacteria: review of current potentials and applications. *Environments* 7(2)13.

Chapter 2

An Evaluation of Fertilizer-Based Disturbances of Agricultural Lands and Their Remediation

Sevinç Yeşilyurt[1,*]
and Muazzez Gürgan[2]
[1]Department of Soil Science and Plant Nutrition, Tekirdağ Namık Kemal University, Tekirdag, Turkey
[2]Department of Biology, Tekirdağ Namık Kemal University, Tekirdağ, Turkey

Abstract

Reduction of fertility is one of the most important problems of the agricultural lands in today's world. This situation is mostly due to the widespread of conventional agriculture to feed the ever-growing population of the world. As a result, unconscious and excessive utilization of fertilizers to obtain the highest yield from unit area led to unfertile and polluted soils. Fertilization performed without proper evaluation of the nutrient element capacity of the soils can cause the micronutrient elements to reach a toxic level and research showed that together with excess phosphorus fertilizer utilization cadmium heavy metal levels have raised. Another source of pollution is the reduction of organic matter in soils below the sufficiency level due to conventional agriculture applications. Organic matter which affects productivity capacity of soils resulted from inaccurate fertilizer applications to eliminate the deficiencies of micronutrient elements led to toxic levels of some micro nutrient elements. The most economical and easy method to apply to remediate the pollution based on fertilizer utilization in the agricultural lands is phytoextraction. This chapter discusses how to form some pollutants due to fertilizers and solutions to this problem.

---

* Corresponding Author's Email: sevincyesilyurt1@gmail.com.

In: Agricultural Soils
Editors: R. Nogueira de Sousa, T. Nascimento Pessoa and K. Ferreira Mendes
ISBN: 979-8-89113-000-5

Keywords: soil productivity, fertilizer, contamination, phytoextraction

Introduction

Agriculture can be defined as the art of obtaining the highest yield from the soil by using plants. Soil is continuously used in order to obtain the highest yield and thus the productivity of the soil should be kept continuous. In order to provide this, it is essential to protect the multidimensional balance and order. For this reason, agriculture can also be defined as the optimum provision of input/output balance in the soil. The subject to be considered is what to provide the soil within order to obtain what. We can succeed as long as we can answer this question well. Because it is deliberate to God to give without obtaining (Sağlam 2012). The agricultural product spectrum changes according to needs and preferences around the world. The highest yields are obtained from cereal, sugar crops and sugar cane agriculture (Figure 1).

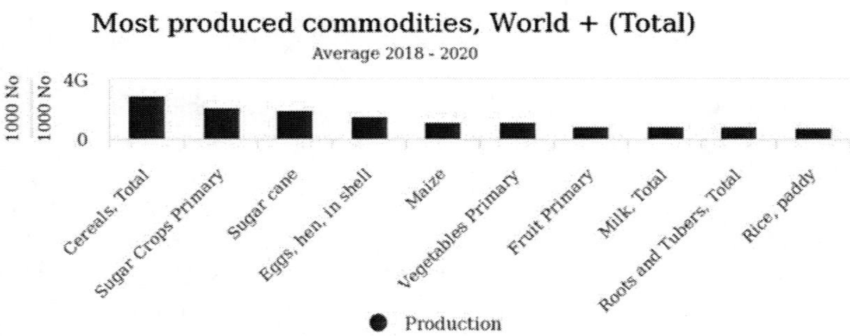

Source: Adapted from: Faostat. (2022).

Figure 1. The main agricultural products around the world in the last three years.

The leading countries for the production of cereals are China, India and Russia and it is Brazil for the production of sugar cane (Figure 2).

Sustainable agriculture is obtaining products of sufficient amount and quality without destruction of natural resources, harm to the ecosystem, and causing environmental pollution (Karaman et al., 2012).

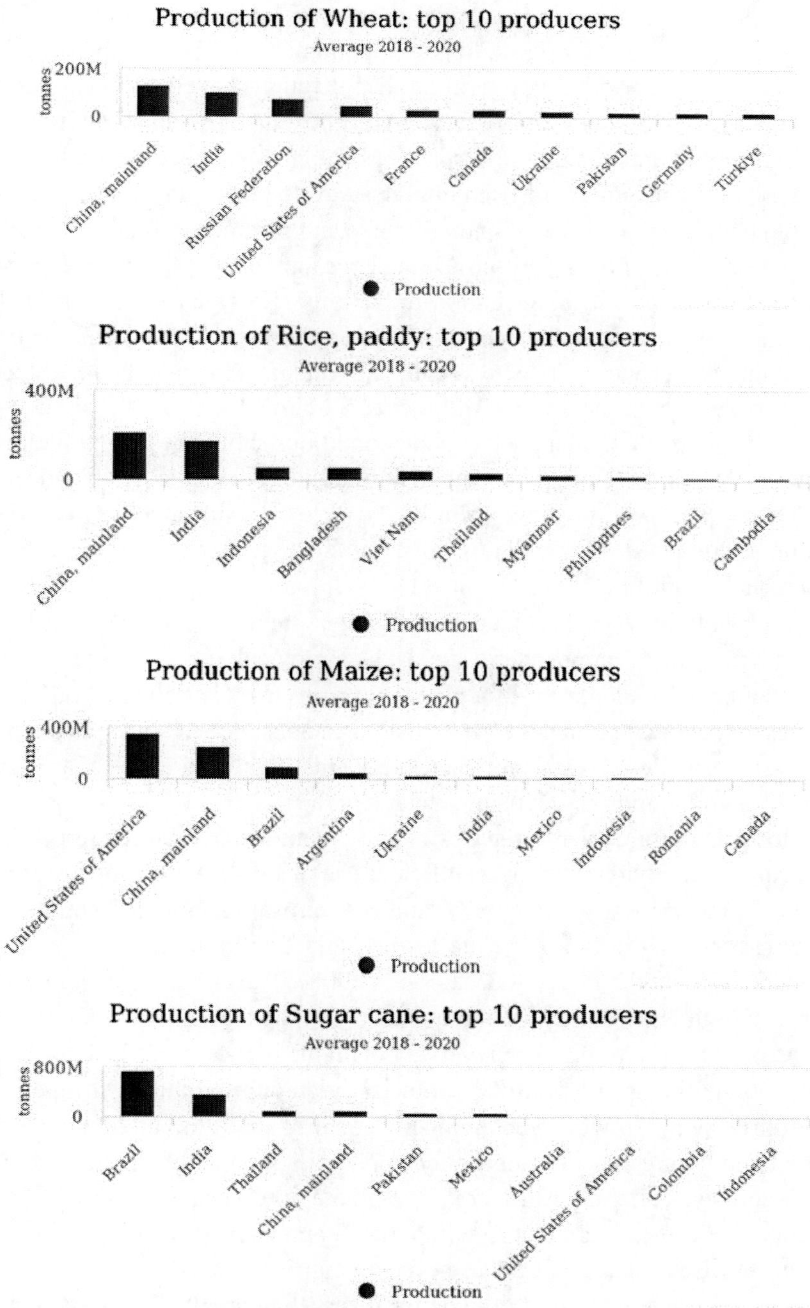

Source: Adapted from: Faostat. (2022).

Figure 2. The leading countries which produce the main food products.

Sustainable agriculture system is a comprehensive system that includes not only environmental effects, but also economic, social, and other subjects. Various different alternative agricultural applications have been mentioned in the concept of sustainable agriculture in recent years. The aims of alternative agriculture systems include restoring the ecological balance which has been disturbed by conventional agricultural systems, formation of a healthier and livable habitat, making use of natural inputs and methods as much as possible instead of artificial ones, soil protecting precautions, increasing soil organic matter, and restore of the lost soil productivity (Karaman et al., 2012). The fundamentals of sustainable agriculture are environmental impact assessment, sustaining soil productivity, and keeping the chemical input at the lowest level to protect the soil's naturality (Adiloğlu and Adiloğlu 2003).

The soil, plant, human, and climate conditions should be in a harmony for sustainable soil productivity and good agriculture. The better this balance, the higher the yield and income, accordingly. When the balance between these factors is collapsed, soil productivity and product yield are negatively affected (Marschner 2008).

Nutrients may be provided to the soil system by chemical fertilizers, organic fertilizers, atmospheric deposits, biological nitrogen fixation, and sedimentation, while on the other hand, they are lost from the soil due to the harvest, leaching, loss in gas form, erosion, and also due to the applications to keep the agriculture sustainable and the yields high (Sağlam 2012). Hence, there is a continuous and dynamic change in the system.

In agricultural systems, there is a strong relationship between fertilization and soil productivity. Moreover, efficient use of fertilizers is quite important to obtain the highest yield, quality, and economical income, to keep the risk of environmental pollution at the lowest, and for the sustainability of soil productivity (Havlin et al., 2005, Roberts 2008). The aim of fertilization is to obtain a high level of quality and quantity. In other words, the main thought of the producer is to follow the way to obtain the highest benefit. This is only possible via the use of fertilizers in the right amount, at the right time in an appropriate way. However, obtaining high quality, high amount of product and high benefits are not only dependent on fertilization. Besides fertilization, irrigation, agricultural pest control, soil tillage, mechanization, seeds of high quality, and ecological factors are important parameters. Usually, around 50% of the fertilizers are expected to be beneficial in terms of soil productivity during the mineralization of organic fertilizers (Sağlam 2012, Adiloğlu et al., 2021).

Chemical Properties of Soil and Productivity

Soil Reaction

Soil reaction is a term used to express the acidity, alkalinity, or neutrality of any soil. Soil reaction appears as a function of concentrations of hydrogen ions (H^+) and hydroxyl ions (OH^-) in soil solution. The concentration of hydrogen ions is a measure of soil acidity while that of hydroxyl ions is the measure of soil alkalinity. Soil reaction is an indicator of numerous soil properties. The appropriateness of any soil as a plant growth medium and its desirability by microorganisms depends on whether soil is acidic, alkaline, or neutral. The amount of amendment matter to change the soil reaction is closely related to the buffer capacity of the soil. The higher the soil buffering capacity, the higher the sulfur and lime amount to change the soil pH. Soil reaction is an important parameter that affects the solubility of all nutrients, their mobility and availability (Adiloğlu and Adiloğlu 2005). That is why the availability and so uptake ability of the nutrients are closely related to the changes of soil pH.

Fertilizers that have organic matter in your composition may increase the buffer capacity of the soil by the addition of organic matter. Hence, the soil can be protected against sudden pH changes and prevent the plants from possible harm.

Cation Exchange Capacity of Soil

Colloidal soil properties direct several chemical reactions and processes going on in the soil system. Colloidal matters are small particles having radius of 0.001-1 m (Bohn et al., 1985). These particles do not precipitate in suspension and also do not distribute homogeneously in the solution. Special processes are required to precipitate them. Colloidal particles that have some optical properties are not visible under simple microscopes. Nutrient that are captured by colloidal fraction may enter the soil solution depending on the need from time to time. Therefore, nutrients are maintained continuously throughout the whole plant development with the help of colloids. That is why not only the concentration of nutrients in soil solution, but also the concentration found in the colloidal fraction in a bounded state is important for the determination of yield strength.

The cation exchange capacity (CEC) of the soil significantly increases upon organic matter addition (Lowenfels and Lewis, 2010). Manure increases the organic matter and humus content of the soil. The CEC of humus and organic matter are also high. Therefore, the soil reserves more nutrient matter, and the washout of nutrients decreases, the buffering of soil increases and the soil becomes more suitable for plant production.

Physical Properties of Soils and Productivity

Organic fertilizers increase the water retention of soil. The effects of the amount of water on the product is significant in lands with low rainfall. In such circumstances, application of manure to the soil increases the water retention and hence the yield. Manure fertilizer increases the stability of the aggragates and changes the pore size distribution. Wortman and Shapiro (2007) found that manure fertilizer increased the number of macro aggregate twice. By this way, the application of manure fertilizer increases the water retention. Moreover, organic fertilizers prevent the water run off the surface, evaporation of water and removal of soil, and increase the water permeability. This is especially important for our country where erosion has largely harmed the soil (Mengel and Kirkby 2001).

Mellowness of Soil and Aggregation

The perfect timing of mellowing of soil is very important in tillage. Problems may arise in production due to the delayed plow and sowing because of late mellowing of the soil. Organic fertilizers provide an ease in the mellowing of soil. Chouichom and Yamao (2011) reported that upon organic fertilizer application, the soil becomes more mellow. Manure gathers the soil particles together in sandy soils and provides aggregation, and increase the water retention capacity. It enters in between the soil particles in clay soil and provides the passing of air and water easily and increases the porosity. As a result, it prepares a more suitable medium for plant production in both kinds of soil (Karaman et al., 2021).

Soil Aeration

Organic fertilizers have positive effect on aeration of soil due to their structure. Especially in heavy clay loam soils, disordered aeration circumstances can be fixed by addition of organic fertilizer. On the other hand, CO_2 and organic acids formed as a result of decay of manure in soil helps the nutrients become available. Usually, the CO_2 level in soils with organic fertilizer applied is higher than that without organic fertilizer (Adiloğlu and Adiloğlu 2009). Organic fertilizers may increase the porosity of soil together with water permeability and hydraulic conductivity (Hussain et al., 2001). The increase in the porosity of soil increases the aeration of soil, and contribute to the soil productivity together with other factors such as increased water content.

Soil Temperature

Temperature retaining capacity of dark colored soils are higher than light colored ones. Soils become darker in color due to use of organic fertilizer applications for long years and therefore their temperature becomes more available for plant development (Sağlam 2012). Moreover, application of organic fertilizers can decrease the evaporation of water from soil and contributes the increase of soil temperature. Also, as the dry weight in the organic fertilizer, the soil temperature increases (Deguchi et al., 2009).

Biological Properties of Soils and Productivity

Soil Microorganisms

The number of bacteria found in soil changes depending on the circumstances and it sometimes may reach up to a hundred million in 1 gr of soil. The most significant reason for bacterial number being higher than other microorganisms in soil is that there are various bacterial species living under different circumstances such as aeration, soil pH, etc. Organic fertilizers have positive impact on both microorganism number and functions. Because manure brings the soil to an optimum situation such as pH, temperature, aeration, and nutrition required by microorganisms. Together with manure, a lot of microorganisms are introduced to the soil. Hence, the biological activity in soil significantly increases (Tok 1997). The study by Bai et al., (2002) showed that soil acidification changed upon application of chemical fertilizers which include nitrogen from ammonium and phosphorus in long term. They reported that ammonium changed the bacterial community and phosphorus decreased fungal diversity. Their findings suggested that chemical fertilizer usage should be replaced by organic fertilizers to improve beneficial microbiota.

Fertilizers and Types of Fertilizers

Organic or inorganic matters applied to soil to increase yield are called fertilizers. Fertilization is the process to apply fertilizers. There are two aims of fertilization. First is enrichment of soil nutrient elements, and the second is enhancement soil physical, chemical and biological properties of the soil for better plant development. Organic fertilizers serve for both and enhance soil physical, chemical and biological properties and the soil nutrient elements. Chemical or commercial fertilizers, on the other hand, only enhances soil nutrient elements. Fertilizers are classified into two classes as organic and chemical fertilizers as a consensus of many researchers. This classification fertilizers having organic nutrient elements are classified as organic fertilizers. For example, manure, green fertilizer, bone powder, blood powder, etc. On the other hand, fertilizers obtained from chemical methods and include one or more nutrient elements are classified as chemical fertilizers. For example, ammonium nitrate, ammonium sulfate, potassium nitrate, calcium nitrate, triple superphosphate, diammonium phospate are chemical fertilizers. Their consumption amount among the world countries is given in Figure 3. Fertilizers can be classified as production method as: Natural fertilizers or artificial fertilizers. Fertilizers that are found naturally or slightly treated fertilizers (Rock phosphate, Chilean nitrate) can be given as examples to natural fertilizers. example to artificial fertilizers are the ones obtained from different fabrication processes (Ammonium sulphate, super phosphate, etc.)

The fertilizers can be classified according to nutrient elements as follows: Fertilizers having only one kind of nutrient element or the ones with more than one nutrient element. In other words, they can be expressed as; (i) Mixed fertilizers: physical mixing of fertilizers which have one or more nutrient elements; (ii) Complex fertilizers: chemical combination of fertilizers having different kinds of nutrient elements (ammonium phosphate etc.) (Adiloğlu and Adiloğlu 2005)

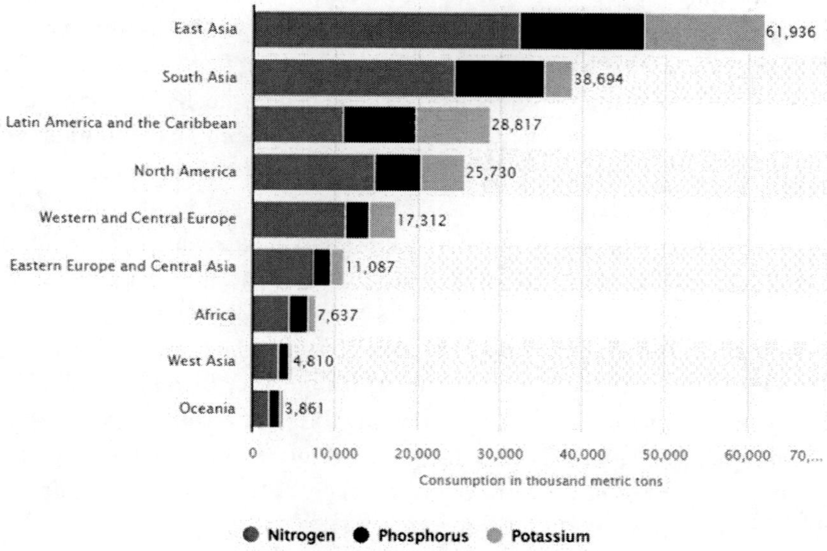

Figure 3. Consumption of main fertilizers worldwide in 2021. (https://www.statista.com).

- ✓ Fertilizers based on physical conditions:
  - Solid fertilizers (crystal, powder, granulated) (e.g., Liquid fertilizers (solution, suspension) (e.g., Urea and ammonium nitrate mixture (Falls and Siegel, 2005))
  - Gaseous fertilizers (liquid ammonium under pressure)
- ✓ Fertilizers based on efficiency
  - Fast release (having higher solubility)
  - Slow release (having lower solubility) (Sağlam 2012; Karaman et al., 2012).

The factors decreasing the efficiency of fertilizers are given in Table 1.

Fertilizer and Pollution

Fertilizer application to soil is known to be carried out since ancient times. The first humans feeding on the natural nutrients ready in natura did not have an effort to use fertilizers. The first efforts in this subject were most probably during the times of people transiting to settled life (nomadic lifestyle) and they started to use fertilizers in either consciously or unconsciously. The first fertilizer used were organic fertilizers. It was realized that yields were higher in places where manure were piled up and hence the fertilization begun. Developments in soil productivity and fertilization were made in middle age and new age. In twentieth century, fertilizer industry developed even more and entered an intense production era. Parallel to fertilizer production, several trials were carried out and important results were obtained about usability and efficiencies of fertilizers (Adiloğlu 2020, Gürgan and Adiloğlu 2021). Development in chemistry industry in the beginning of twentieth century and during war times reflected to the fertilizer industry and various fertilizers were produced. There are still various solid and liquid fertilizers available. The development will not seem to stop and many different fertilizers will continue to be produced due to the fact that there is an effort to obtain higher yield from unit area. There are some factors affecting the yield obtained from the unit area. They are irrigation, fertilization, using seeds of quality, tillage, agricultural pest control, mechanization, environmental factors, etc. They are effective on the yield one by one. However, it is accepted that the role of fertilization is equal to the others all together. On the other hand, insufficient fertilization results in loss of product while excess and unconscious fertilization results in waste of sources (Lax 1991, Lewan 1994, Sağlam 2012, Adiloğlu and Göker 2021, Adiloğlu 2021).

Table 1. Factors decreasing fertilizer efficiency. Source: Adapted from: (Karaman and Turan 2012)

Factors	Ratio of decreasing efficiency (%)
Improper drill	10-20
Improper plant	20-40
Late sewing and planting	20-40
Early sewing and planting	5-20
Insufficient vegetation	10-25
Insufficient watering	10-20
Diseases	15-20
Pests	5-50
Unconscious fertilizing	20-50
False timig for fertilizing	5-10

Heavy Metals and Relationship with Fertilizers

Heavy metals such as Hg^{2+}, Cu^{2+}, Cd^{2+}, Ni^{2+}, Zn^{2+}, Pb^{2+} replace with Mg in the chlorophyll and cause the distortion of structure (Kupper et al., 1998). Chlorophyll is one of the most important parameters in photosynthesis. The central atom in the chlorophyll structure is Mg. If heavy metals replace with this element in chlorophyll, the change in the structure of chlorophyll decreases the efficiency of photosynthesis. Hence, plant growth and development decreases. Excess and unconscious fertilization may result in cadmium pollution (Karaman et al., 2012, Adiloğlu. 2016).

Iron toxicity is usually observed in water saturated soils. Even though the critical toxicity level was determined to be 500 mg kg-1, it may change depending on the iron nutrition of plants and other factors. Iron toxicity is important not only in water satisfied soils, but also under arid conditions (Price and Hendry 1991, Aydın and Turan 2002). In the case of excess iron, symptoms appear similar to phosphorus deficiency. In the plants grown in water satisfied conditions such as rice, high Fe uptake can cause toxicity. Tanning of rice leaves is a typical Fe toxicity. In this case, Fe content of leaves can be 700 mg kg^{-1} or higher (Yamauchi 1989). In case of iron toxicity, tanning is a result of oxidized polyphenols (Yamauchi 1993). Potassium application can be done in order to eliminate iron toxicity. In this case, K decreases Fe uptake and increases oxidation potential of roots (Güneş et al., 2000, Adiloğlu and Gürgan 2021).

Zinc toxicity leads to lower the growth of root and shoot growth, thin the roots, coil the young leaves, and causes chlorosis. In case of excess of zinc, cell growth and elongation are inhibited, cell organelles are ruptured and chlorophyll synthesis slows down. (Rout and Das 2003). Excess zinc makes the root meristem cells lignify and both root and shoot growth are inhibited (El-Ghamery et al., 2003). A high level of zinc also inhibits the plant's benefit of iron and replace the Mg in the center of chlorophyll (Van Assche and Clisters 1990).

Mn toxicity occur in cases of high soil acidity and especially when Ca and Mg are uptake less by the plants. This situation is more significant in cabbage, cauliflower, and barley. The diagnosis of Mn toxicity is not as easy as Mn deficiency. Leaves turn yellow, dots appear on the sides of leaves, and plants do not grow long in the case of manganese toxicity. The most sensitive plants to manganese toxicity are tobacco, clover, trefoil, vegetables, and citrus plants. Irregular chlorotic spots appear between veins, then purple, dark brown or black necrotic stains form. Stains are intense on the sides of the leaves and leaves can coil up to the center (Adiloğlu 2017).

Copper toxicity level differs between plants (Hodenberg and Finck 1975, Robson and Reuter 1981). For example, bean is more tolerant to copper than corn. In case of excess copper, root growth is affected more since more copper is accumulated in roots than aerial parts. Plants with no copper tolerance, the first symptoms of copper toxicity are root growth decrease, K secretion and harm in plasma membrane (Baker and Walker 1989, De Vos et al., 1991). Copper toxicity sometimes cause Fe deficiency (Taylor and Foy 1985, Bergmann 1992, Adiloğlu 2017). Decreases in IAA oxidase activity in plants exposed to high Cu concentration reveal decrease in root growth or formation of side roots.

Easy and Cheap Remediation of Pollution

Phytoremediation concept is a general name given to the technology which uses plants to remediate pollution. The aim in the process of remediation of heavy metals using plants is the conversion of heavy metals captured by the soil into a more controllable and transferable form. For this reason, phytoremediation method should not be considered as a final remediation removal method. Final removal or remediation would be succeeded with burning the plants or storage of plants in an appropriate place after phytoremediation.

Hyperaccumulator plants of 400 kinds are known to accumulate metals and they are known to belong *Asteraceae, Brassicaceae, Caryophyllaceae, Cyperaceae, Cunouniaceae, Fabaceae, Flacourtiaceae, Lamiaceae, Poaceae, Violaceae, and Euphobiaceae. Brassicaceae* families (EPA 2004, Adiloğlu and Gürgan 2020).

Herbal remediation converts metals and radioactive elements into transferable chemical forms and make them immobile. They become the less risky form for human and environmental health. When compared with the other remediation methods, being cost efficient is one of the superior aspects of this method. A more favorable medium is being formed for the other plants after phytoremediation (Köseoğlu 2007).

The most appropriate plant for this method should have a strong root system, should accumulate high level of metals in harvestable parts, be fast growing and promote high amount of biomass in the land. Hyperaccumulators are plants that can grow in the presence of concentrations of some elements that can harm many plants (Watanabe 1997, Reeves and Baker 2000, Adiloğlu and Pamay 2021).

The accumulation of heavy metals in soil affects soil production potential and ecosystem, as well as animal and human health via food chain. The high expenses of cleaning classical removal technologies limit the applications of them. Therefore, there is a need for new, highly efficient and cost-effective technologies. Remediation using plants was therefore developed to be more cost-effective and more aesthetic than the other cleaning methods.

According to Salt et al. (1998), phytoremediation techniques are divided into subclasses such as phytoextraction, phytodegradation, rhizofiltration, phytostabilization, phytovolatilization, and rhizodegradation. Phytoremediation techniques are effective on remediation of low to medium degree of polluted areas. On the other hand, this remediation technique can be used together with other remediation methods as a complementary method (Türkoğlu 2006).

Conclusion

The continuous decrease of the agricultural soil productivity should be inhibited by the applications of chemical, organic and biofertilizers. However, unconscious fertilizing and wrong applications negatively affect the soil quality and healthy agricultural production. Therefore, fertilizer spectra should be enlarged for the purpose of healthy agricultural productions and to enhance soil productivity. Single way of chemical fertilizer applications can result in contamination of soils with heavy metals. That is why organic and biofertilizers should be brought into prominence. Studies on this area should be supported and their numbers should be increased. Moreover, these studies should enlighten the conditions and bring solutions to the fertilization caused problems all over the world.

References

Adiloğlu, A. and S. Adiloğlu. 2003. The investigation of some available trace element contents of acid soils in Turkey. *Archives of Agronomy and Soil Science* 49 (2): 179-185.

Adiloğlu, A. and S. Adiloğlu. 2005. Fertilizer Consumption in Turkey Between 1960 and 2002. *Bulgarian Journal of Agricultural Sciences* 11(3): 317-321.

Adiloğlu, S. and A. Adiloğlu. 2009. The Effect of Increasing Nitrogen and Zinc Application on the Nickel, Lead and Cadmium Contents of Maize Plant in Typic Xerochrept and Calcixeroll Soils. *Asian Journal of Chemistry* 21(2): 1475- 1482.

Adiloğlu, S. 2016. Using Phytoremediation with Canola to Remove Cobalt from Agricultural Soils. *Polish Journal of Environmental Studies* 25(6); 2251-2254.

Adiloğlu, S. 2017. Interaction of Manganese and Some Heavy Metals in Dock (*Rumex patientia* L.) Plant for Remediation of Contaminated Soils. *Desalination and Water Treatment* 93: 335-338.

Adiloğlu, S. and M. Gürgan. 2020. Karoline Møller (Editor) Book title: *Brassica napus: Cultivation and Uses.* Chapter title: Brassica napus: Phytoremediation. Nova Science Publishers, Inc. pp. 134.

Adiloğlu, S. 2021. Relation of Chelated Iron (EDDHA-Fe) Applications with Iron Accumulation and Some Plant Nutrient Elements in Basil (*Ocimum Basilicum* L.). *Polish Journal of Environmental Studies* 30(4):3471–3479.

Adiloğlu, S. and M. Göker. 2021. Phytoremediation: Elimination of Hexavalent Chromium Heavy Metal Using Corn (*Zea Mays* L.). *Cereal Research Communications* 49 (1): 65-72.

Adiloğlu, S. and M. Gurgan. 2021. Book title: *Agricultural Soil: Management and Impacts on the Environment.* Chapter title: "Remediation Of Boron Toxicity Using Bacteria" Editor: Lynn Mcconnell, Nova Science Publishers, Inc.. Nova Science Publishers, Inc.

Adiloğlu, S. and S. Pamay. 2021. James C. Flores (Editor) Book title: *The Future of Phytoremediation.* Chapter title: Phytoremediation and Hyperacumulative Families. Nova Science Publishers, Inc. pp. 305.

Adiloğlu, S., Eryilmaz Açıkgöz F., Bellitürk K., Gürgan M. Solmaz Y. and A. Adiloğlu. 2021. The effects of increasing amounts of vermicompost and a fixed amount of *Rhodobacter capsulatus* applications on macro and micro elements of plant and soil samples *Journal of Plant Nutrition* 44(19): 2876-2884.

Adiloğlu S. 2021. Relation of Chelated Iron (EDDHA-Fe) Applications with Iron Accumulation and Some Plant Nutrient Elements in Basil (*Ocimum Basilicum* L.). *Polish Journal of Environmental Studies* 30(4):3471–3479.

Baker, A. J. M. and P. L. Walker. 1989. Physological response of the plants to heavy metals and the quantification of tolerance and toxicity. *Chemical Speciation & Bioavailability* 1,7-17.

Bai, Y C. Chang, Y Y. Hussain, M. Lu, B. Zhang, J P. Song, XB. Lei, XS.and D. Pei. 2020. Soil Chemical and Microbiological Properties are Changed by Long-Term Chemical Fertilizers that Limit Ecosystem Functioning. *Microorganisms.* 8(5):694.

Bergmann, W. 1992. *Nutritional Disorders of Plants.* Gustav Fisher Verlag Jena. Stutgart-New York. pp. 741.

Chouichom, S. and M. Yamao. 2011. Organic fertilizer use in northeastern Thailand: an analysis of some factors affecting farmers' attitudes. In *Sustainable agricultural development* Springer, Dordrecht. pp. 185-196.

De Vos, C. H. R., Schat H., de Waal, M. A. M., Woojis, R. and W. H. O. Ernst 1991. Increased resistance of copper induced damage of the roots cell plasmalemma in copper tolerant *Silene cucubalus. Physiologia Planttarum* 82:523-528.

El-Ghamery, A. A., El-Kholy, M. A and El- Yousser, A. 2003. Evaluation of cytological effects of Zn^{2+} in relation to germination and root growth of *Nigella sativa* L. and *Triticum aestivum* L. *Mutation Research* 537:29-41.

EPA, (Environmental Protection Agency) 2004. *Radionuclide Biological Remediation Resource Guide.* EPA 905-B-04-001, U.S. Environmental Protection Agency, Region 5 Superfund Division Chicago, Illinois 60604 USA.

Falls, J. H., S. A. Siegel. 2005. Fertilizers, Editor(s): Paul Worsfold, Alan Townshend, Colin Poole, *Encyclopedia of Analytical Science (Second Edition)*, Elsevier, pp 1-8, Agriculture Faculty, No:1514.

Güneş, A., Alpaslan, M. and A. İnal 2000. *Plant nutrition and fertilizer* A.U.

Gürgan M., and S. Adiloğlu. 2021. The Effects of Increasing Concentrations of Iron Fertilizer on Antibacterial Activity of Basil (*Ocimum Basilicum* L.). *Industrial Crops & Products* 170: 113768.

Havlin, J. L., Beaton, J. D., Tisdale, S.L. and W. L. Nelson. 2005. *Soil Fertility and Nutrient Management: An Introduction to Nutrient Management.* 7th Edition. 515 p. Pearson/Prentice Hall. Upper Saddle River, NY Roberts.

Hodenberg, V. A. and A. Finck. 1975. Ermittlung von toxizitaitsgrenzwerten fur zink, kupfer und blei in hafer und rotklee. [Determination of toxicity limits for zinc, copper and lead in oats and red clover.] *Zeitschrift für Pflanzenernährung und Bodenkunde* [*Journal of Plant Nutrition and Soil Science*] 4(5):489- 503.

Hussain, N. Hassan, G. Arshadullah, M. and Mujeeb. F. 2001. Evaluation of Amendments for the Improvement of Physical Properties of Sodic Soil. *Internatıonal Journal of Agrıculture & Bıology* 3(3):319-322.

Karaman, M. R., Aydın A., R Brohi, A., Güneş, A., İnal, M., Kaplan, V., Katkat, Korkmaz, A. 2012. *Plant Nutrition.* Ankara: Dumat Ofset.

Karaman M. R. and M. Turan. 2012. Sustainable management strategies in plant nutrition and fertilizer efficiency parameters. *Soil Water Journal*, 1(1): 15-21.

Köseoğlu, C. 2007. Applicability of phytoremediation in waste sludge stabilization, Çukurova University, *graduate thesis of the institute of science*, Adana.

Lax, A. 1991. Cation exchange capacity induced in calcareous soils by fertilization with manure. *Soil Science* 151:174-178.

Lewan, E. 1994. Effects of a catch crop on leaching of nitrogen from a sandy soil. *Plant and Soil* 166:137-152.

Lowenfels, J. and Lewis. W. 2010. *Teaming with Microbes: The Organic Gardener's Guide to the Soil Food Web*, Timber Press.

Marschner, H. 2008. *Mineral Nutrition of Higher Plants*. Digital Print. Academic Press., pp. 889.

Mengel, K. and E.A. Kirkby. 2001. *Principles of Plant Nutrition. 5th Edition.* Kluwer Academic Publishers. ISBN: 1-4020-0008-1, Dordrecht, The Netherlands.

Reeves, R. D. and A. J. M. Baker. 2000. Metal– Accumulating Plants. in: Raskin, I. and Ensley, B.D., Eds. *Phytoremediation of Toxic Metals: Using Plants to Clean–Up the Environment*. New York, John Wiley and Sons, p. 193–230.

Robson, A. D. and D. J. Reuter. 1981. Diagnosis of copper deficiency and toxicity. In: *Copper in soils and plants* (J. F. Loneragan, A. D. Robson and R. D. Graham, eds.) 287-312 Academıc Press, London.

Rout, G. R. and P. Das. 2003. Effect of metal toxicity on plant growth and metablism: I. Zinc. *Agronomie* 23:3-11.

Sağlam, M. T. 2012. Soil chemistry. *Tekirdağ Namık Kemal University lesson book, No.1* Tekirdağ.

Salt, D. E., Blaylock, M., Kumar, P. B. A. N., Dushenkov, V., Ensley, B. D., Chet, L. and L. Raskin. 1995. Phytoremediation: A Novel Strategy for the removel of Toxic Metals from the Environmental Using Plants. *Nature Biotechnology* 13: 468-474.

Shin, D. Hidenori, K. Osamu, T. Akihide, F. and U. Sunao. 2009. Compost application increases the soil temperature on bare Andosol in a cool climate region. *Soil Science and Plant Nutrition* 55(6): 778-782.

Taylor, G. J. and C. D. Foy. 1985. Differential uptake and toxicity of ionic and chelated copper in *Triticum aestivum. Canadian Journal of Botany* 63:1271-1275.

Tok H H. 1997. *Soil Biology*. T. Ü. Tekirdağ Agriculture Faculty, Class book. Tekirdağ.

Türkoğlu, B. 2006. *Soil pollution and remediation of polluted soils.* Çukurova University, graduate thesis of the institute of science, Adana.

Van Assche, F. V. and H. Clijsters. 1990. Effects of metals on enzyme activity in plants. *Plant, Cell Environment* 13:95-206.

Watanabe, M. E. 1997. Phtoremediation on the brink of commercialization. *Environmental Science and Technology* 31: 182-186.

Wortmann, S. 2007. The Effects of Manure Application on Soil Aggregation C. S. Wortmann Æ C. A. Shapiro *Nutrient Cycling in Agroecosystems.*

Chapter 3

Conventional and Alternative Nutrient Sources for Soil Fertility and Strategies to Improve Their Use Efficiency

Ana Carolina Benites Aquino*
Rodrigo Nogueira de Sousa
Deyvison de Asevedo Soares
and Caroline de Figueiredo Oliveira
Soil Science Department, University of São Paulo, Piracicaba, São Paulo, Brazil

Abstract

Mineral fertilizers are readily soluble and available nutrients in the soil solution for plants uptake. These sources are necessary for food production to remain constant. However, the manufacturing process of conventional nutrient sources is surrounded by environmental problems. Fertilizers are still produced on a large scale, but, long-term mineral extraction can lead to a shortage of nutrient sources. Soil, in its complexity and heterogeneity, is the basis of agriculture. In addition to the problems associated with the production offertilizers, their use on soil can lead to low fertility over time by affecting its organic content. The physical, chemical, and biological quality of the soil must be maintained tosustain production. The biological attributes as organic matter and microorganisms, when affected, also affect chemical and physical attributes. Organic matrix fertilizers can be the key to increasingly sustainable agriculture and producing quality agricultural commodities. The disadvantage of organic fertilizers is the need to apply high doses to

* Corresponding Author's Email: ana.aquino@usp.br

In: Agricultural Soils
Editors: R. Nogueira de Sousa, T. Nascimento Pessoa and K. Ferreira Mendes
ISBN: 979-8-89113-000-5

meet crop requirements. However, when organic fertilizers are combined with minerals, know as organo-mineral fertilizers, the amount of fertilizer needed to be applied is reduced, making application advantageous. Thus, the authors' objective in this chapter is to present the potential environmental problems of NPK fertilizer production and how the use of organic residues can be beneficial by providing an environmentally sound destination for the residues and reducing dependence on mineral fertilizer imports from other countries.

Keywords: sustainability, production, food, agriculture

Introduction

To supply the food demand of an expanding population is a challenge that humanity has been carrying throughout the years. The world population will have about 9.7 billion people in 2050, with a growth rate of more than 20% (2022-2050), according to ONU (2022). In order to meet the goal of feeding this population and producing more food it is necessary to increase the productivity of crops in the areas already cultivated, without the need to clear new areas of natural vegetation.

The domestic consumer culture of a population influences policy decisions related to the international agricultural sector and can dictate the march of food production, the nutritional content of products, and the means and techniques employed in commodity and food production, as they supply different consumer markets.

Agricultural policies could favor more ecological, sustainable, and resilient approaches, with the adoption of technologies that enhance the use of agricultural inputs and minimize their environmental impacts from manufacture to use. The use of land according to agricultural aptitude results in financial gains and conservation of these areas and natural resources. Tax incentives and subsidies on the preservation, restoration and productivity of agricultural areas can be an incentive to make the systems more sustainable and think about long-term productivity.

Fertilizers are a critical point for the food production chain. The efficient use of these agricultural inputs affectsthe quality and quantity of food produced worldwide. Mineral fertilizers are a criticalfactor in increasing crop yields. On the other hand, the production of these fertilizers can cause environment damage, such as the emission of greenhouse gases. The sustainability of the agricultural fertilizer chain is not only related to the production of these inputs, but also to the use of residue products from other industries, that can be used as potential fertilizers.

Industrial, animal, and plant residues can be processed and used as organic fertilizers in agriculture. However, the need to apply high doses of these residues in agriculture is a factor that discourages their use. The combination of residues with mineral fertilizers is an option that offers a sustainable destination for the residues while the use of mineral fertilizers is lower. Thus, the application of this product to crops is feasible and necessary in smaller doses, when compared to the application of organic sources. Thus, the aim of the authors with this chapter is to present the possible environmental problems arising from the production of mineral NPK fertilizers and how the use of organic residues can be beneficial by offering an environmentally correct destination for the residues and reducing the dependence on the importation of mineral fertilizers from other countries.

Conventional Fertilizers

Nitrogen Fertilizers

Nitrogen (N) is a mobile element in the soil and is dynamic in the soil-plant-atmosphere system. It is required in large quantities by agricultural crops and its supply via mineral fertilizers to meet crop needs is what ensures modern agriculture in the current model. Plants with the ability to fix nitrogen in the soil can be used in some cropping models. However, for large-scale agriculture requires fertilizers that facilitate crop rotation with crops of greater agricultural interest (Robertson and Vitousek, 2009).

Nitrogen within the agricultural system promotes changes in the microbiota, which alters the rates of decomposition and mineralization of organic matter, and processes such as nitrification, denitrification, and methane oxidation, in addition to affecting on trophic relationships, with increased plant resistance to pests. The N mobilized by plants is mostly reactive and is present in biologically active forms in the atmosphere. Reactive nitrogen can lead to the loss of biodiversity, besides influencing the quality of the air, as one of the greenhouse gases, and of the water, through contamination of the water table and eutrophication.

In the biosphere, N_2 corresponds to 79% of the composition of dry air, with a triple bond between molecules. This bond requires the energy of

945 kJ mol^{-1} to be broken, understood as extremely strong, and can be broken by atmospheric electrical discharge (in small amounts) and by microorganisms, with emphasis on biological N_2 fixers, which convert the inert element to the form of NH_3, reactive (Geng et al., 2019). The main cereals consumed by humans (corn, rice, and wheat) are produced using about 50% of the total nitrogen fertilizer produced worldwide, while P and K account for 49% of consumption. With the large world production of soybeans, oilseed crops account for 7% of nitrogen fertilizer consumption, and P and K account for 15 and 20%, respectively (Heffer and Prud'homme, 2016).

In the production of nitrogen fertilizers, this bond is broken with heat and pressure in a process known as Haber-Bosch synthesis, which is responsible for the additional of about 120 Tg of N $year^{-1}$ when fertilizers are applied through the soil (IFA, 2016). In this process, the combustion of natural gas (methane - CH_4) leads to the production of hydrogen, which reacts with N_2 under conditions of high temperature and pressure, to form ammonia (NH_3), which represents more than 1% of the world's energy consumption (Engin Sagirli et al., 2021). To produce 1 metric ton of nitrogen fertilizer by this process, the consumption of natural gas is estimated to be 873 m^{-3}. In addition to natural gas, petroleum can also be used in the process.

Even today, with technological advances, there are some points difficult to measure when it comes to nitrogen: i) measuring the content of nitrogen in the soil, given the great dynamism of the nutrient, ii) the amount of the element that is deposited in the soil by microorganisms, through the Biological Nitrogen Fixation (BNF) and, iii) determining the losses of the nutrient in the soil-plant system. N applied to the soil via fertilizers suffers losses to the environment, with negative consequences (Table 1). The application in large quantities of nitrogen fertilizers in reactive forms (ammonium, nitrate and oxides), results in volatilization, leaching, and gas emissions into the atmosphere (Ladha et al., 2020). The emission of nitrous oxide (N_2O) into the atmosphere since the Industrial Revolution has increased by 20% through the use of nitrogen fertilizers. With the intensification of human action, the amount of nitrogen released into the atmosphere has been greater compared to carbon, generating concern that this element may become the "new carbon" and a potential aggravator of global warming (Battye et al., 2017).

The fertilizer market is volatile and buyers are hostage to those who produce the fertilizers. The energy prices in some countries, especially developing ones, are a limiting point that disfavors the production of nitrogen fertilizers and favors the importation of the product from other countries, making the national agricultural market vulnerable to exchange variations.

Besides this, the climate targets that affect the production of inputs in these countries are not elaborated in such a way as to restrict and inhibit over-application and environmental problems (Jayne et al., 2018).

The use of nitrogen fertilizers varies according to the region and the countries' financial and environmental policies. Underdeveloped countries have fewer resources compared to developed countries. Thus, they invest in other aspects of the production chain to achieve higher yields with fewer resources, such as genetic improvement. Thus, the use of fertilizers is lower in these countries. Taking into account international policies, the Sustainable Development Goals (SDGs), defined by the United Nations Development Programme (UNDP) in 2016, seek the optimization of nitrogen flows in the soil-plant-atmosphere system. The 17 SDGs address hunger, poverty, climate, hygiene, and resource-conscious consumption, taking into account the economic, environmental, and social pillars.

Managing reactive nitrogen applications to the soil is the challenge that remains to meet the advances in the food production chain. One of the most relevant approaches to meet the challenge is to improve Nitrogen Use Efficiency (NUE), which is the result of Nutrient Uptake Efficiency (NAE) and Nutrient Utilization Efficiency (NUtA), while minimizing environmental losses. These fertilizers have low stability in the soil solution and the use efficiency is considered low. For cereal crops, only 51% of the total nitrogen applied to the soil is actually utilized by the plants (Cameron et al., 2013). Techniques associated with nitrogen fertilizers can be used to minimize nutrient losses, as cited in Table 1.

Table 1. Various N forms and their environmental impact

Different forms of N		Environmental impact
NH_3	Ammonia, the primary form of reduced N	Used for fertilizers and emitted into the air
NH_4^+	Ammonium ion	Contributes to particulate matter and water pollution
N_2O	Nitrous oxide	A greenhouse gas. It also depletes the stratospheric "ozone layer"
NO	Nitric oxide, the primary form of oxidized N	Emitted to the air from soils and burning of crop residues
NO_2	Nitrogen oxide	Rapidly formed in the air from NO
NO_x	The mixture of NO and NO_2	Affects human health
NO_3	Nitrate ion	Contributes to particulate matter and water pollution
OrgN	Organic nitrogen	A plethora of forms with many impacts
PM	Particulate matter	Causes air pollution, adversely affecting health and visibility

Source: Ladha et al., (2020).

Among the technologies, we can cite the application of fertilizer in deeper layers, slow and controlled release materials, and partial fertilization of the recommended dose. The use of biological sources, such as legumes and nitrogen-fixing microorganisms, can complement/supplement the N needs of crops, making production attractive by reducing costs with chemical inputs, and minimizing environmental damage.

Phosphate Fertilizers

Phosphorus participates in vital processes in plant metabolism, including photosynthesis, signal transduction, macromolecular biosynthesis, respiration, and chemical energy production, in the forms of ATP and NADPH. The levels of P in the plant tissue vary between 0.2% and 0.4% of the plant's dry weight, amounts considered small when compared with nitrogen, for example (Mullen, 2019). This nutrient, which restricts plant growth and development, is one of the major obstacles to contemporary agriculture. The dynamics of phosphorus in the soil are related to the balance between the organic and inorganic phases and depend on factors such as soil and climatic factors and soil management. Climatic variables influence the rate of solubilization of fertilizers and mineralization of organic matter. Soil characteristics, as mineralogy and texture, determine the type and intensity of chemical reactions that occur after fertilization. The cropping system, soil tillage, chemical, physical and biological characteristics of the fertilizer, in addition to the fertilizer doses applied, and the quantity and quality of plant biomass in the area are the factors related to soil management (Nunes et al., 2020).

The natural source of P majority found in nature for most soils is apatite (mineral - $Ca_{10}(PO_4)_6X_2$) or rock phosphate (RP). Apatite is a compound of calcium and phosphorus that is highly insoluble. In highly weathered soils, there is little or no amount of the minerals that compose apatite. In this mineral, the chemical and biological action lead to rock solubilization and releases the constituent P to soil solution to be used by plants and microorganisms. This P is highly reactive, so it can be incorporated into organic matter or other highly insoluble inorganic minerals such as aluminum (Al) and iron (Fe) in acid soils, and with calcium (Ca) in high pH soils.

In highly weathered soils, P is adsorbed by Al and Fe oxides, compounds of moderate solubility (Figure 1). The solubility of these compounds increases with increasing pH around 6.5. With increasing pH, the formation of highly soluble Ca-P compounds occurs. With increasing pH in ranges above thoseconsidered neutral, the concentrations of soluble P gradually decrease. The low solubility of inorganic P makes it necessary to apply soil fertilizers on a much larger scale than the crops need. As shown in Figure 1, it is possible to observe the forms of P formed with the variation of the pH of the soil. As the pH increases, more of the phosphorus forms change from bonds with acid cations (Fe and Al) to calcium forms.

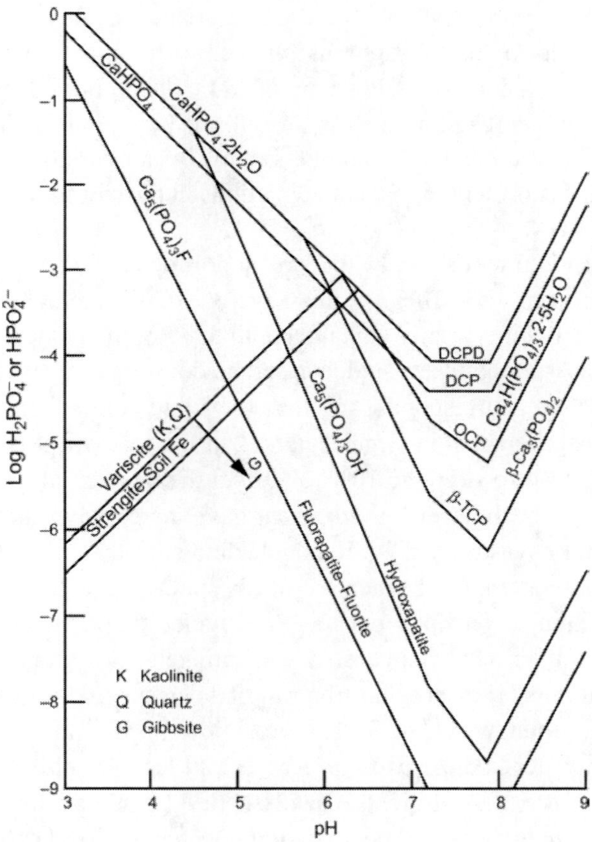

Source: (Mullen, 2019).

Figure 1. Solubility of orthophosphate from variscite ($AlPO_4 \cdots 2H_2O$), strengite ($FePO_4 \cdots 2H_2O$), and various Ca-phosphates as a function of solution pH. These equilibria assume a Ca^{2+} concentration of $10^{-2.5}$ mol L^{-1} and CO_2 at 0.03 kpa. DCPD, calcium phosphate dehydrate, brushite; DCP, calcium phosphate, monetite; OCP, octacalcium phosphate; β-TCP, β-tricalcium phosphate.

The variation of phosphorus in the soil depends on the mineralogy and the degree of weathering. In sandy soils with low organic matter content, the total P content can be in the order of 100 mg kg^{-1} . In less weathered soils rich in primary materials, the clay or organic matter content can be of the order of 2,000 mg kg^{-1}. In general, inorganic P predominates. In his study of P, in weathered soils, Maranguit (2017) found that the P stock was dominated by inorganic forms (60 to 85%) in all land-use types. The organic P content becomes more important as the organic matter content increases. The predominant form of phosphorus available to plants in the soil is orthophosphate, such as $H_2PO_4^-$ or HPO_4^{2-}. This orthophosphate, from fertilizers or mineralized organic matter, forms precipitates with Al-, Fe- or Ca-, or can be adsorbed by Al- and Fe-oxides. Because of this dynamic in the soil, soluble P is present in low concentrations in the soil solution, around 0.1 to 1 mg L^{-1}.

The organic fraction can be the largest portion of the total nutrient in the soil, ranging from 3% to 90%, with an average of 30% to 50% in most soils. Organic matter can typically contain 1% to 3% P in its composition. As soil depth decreases, organic P also decreases, as does organic carbon. This form of the nutrient corresponds to the recycling of animal, plant, and microorganism remains, in larger amounts in the form of phytin phosphates and inositol, corresponding to 10% to 50% of the total organic P.

The concerns about the environment have increased in the last decades. The search for environmentally friendly technologies in agriculture is gaining more and more attention. In the case of phosphorus, the problem can be the over-application of fertilizer over a much wider area than is necessary for crops to meet their nutrient needs. When fertilizers are applied in areas close to surface waters, it is possible for runoff to occur and the nutrient to be deposited in groundwater, or for soil erosion to occur and adsorbed P to be washed into water bodies. These processes can lead to eutrophication of the water bodies, as excess nutrients, usually N or P, accelerate the production of algae, which cloud the water, decrease light penetration, and as a result impair aquatic plant life in the environment. Lower oxygen levels caused by the new environment are harmful to aquatic life because they make the habitat unsuitable.

Therefore, measures that mitigate the negative impacts caused in the ecosystems by the enrichment of the hydrosphere with phosphorus and also nitrogen must be taken, to curb the high social costs of reducing the nutrient levels in waters. And, despite a certain decline of P levels in rivers and lakes, as a result of lower sewage discharges, the annual storage of the nutrient in watersheds remains high, because the export rate of phosphorus in rivers is lower than the inflow of the element in watersheds due to human action. This deposition of P in watersheds is the result of a long period and is difficult to quantify.

Surface runoff, which carries soil particles and also nutrients, occurs in different proportions within the same region, making the dispersion of P dispersed in both terrestrial and aquatic environments. The environmental imbalance then results from the disturbance of what would be the natural cycle of phosphorus, resulting from the incorrect use, distribution, loss, and accumulation of the nutrient, which makes the biosphere sensitive to possible environmental changes, that harm the other living beings present in the habitat.

The lack of governance, oversight, and geographic segregation of the means of production of the agricultural and livestock sectors, coupled with the exponential growth of food commodities as a result of population growth, intensifies the physicochemical processes linked to the phosphorus cycle and thus makes it difficult to make environmental predictions with the unforeseen consequences of this acceleration caused by the global market. As an example of these abrupt changes, we can cite the time when nutrient sources may become scarce due to finite rock reserves and the change from healthy ecosystems to scenarios that harm the life of fauna and flora.

To reverse the negative effects of phosphorus overuse in the systems, residue water treatment, and sustainable cropland management are needed. The P deposit in the soil should be accessed so that finite resources would be reduced. Thus, a smaller amount of fertilizer is deposited in the soil, to increase the levels of the nutrient, which disfavors the chemical balance of reactions and tends to increase the levels retained in the soil.

History shows the enormous evolution of the use of phosphate fertilizer sources in agriculture, with production initially reliant on organic matter from natural sources like manure and human waste (Figure 2). In the mid-20th century, with the increased global demand for food, guano was still the main source of nutrients. However, in the middle of this century, soil fertilization using phosphate rock gained prominence (Cordell et al., 2009).

The fact is that phosphorus is a nutrient required by plants, so it cannot be replaced by any other element in the periodic table. The food production chain requires considerable amounts of this nutrient to keep food production constant. The sources, from which phosphate fertilizers are produced, are non-renewable resources and are expected to be depleted, and the nations that own them have control over these reserves. This behavior has resulted in other countries being held hostage to external sources of the nutrient for food production. Excessive application of the nutrient to the soil increases what we call "legacy P," which causes imbalances in the chemical fraction of the soil and favors the retention of the nutrient, making it unavailable to plants, besides possible environmental damage.

Alternative sources of phosphate fertilizers can be the key to the independence of countries in producing their own fertilizer, or at least, in a first moment, reduce imports and have autonomy in the production of sources for domestic use. It is noteworthy that the use of treated residues as a source of nutrients can reduce fertilizer costs, while offering an environmentally correct destination for the residues, making the activity more sustainable and preserving areas of modern agriculture and ecosystems, as well as the fauna and flora community and the natural characteristics of the areas. The organic matrix of the residues can improve the efficiency of the use of phosphate fertilizers, since it contains in its composition a rich range of microorganisms, that are important for the P cycle and for increasing the availability of the nutrient in the soil solution, to be absorbed by plants.

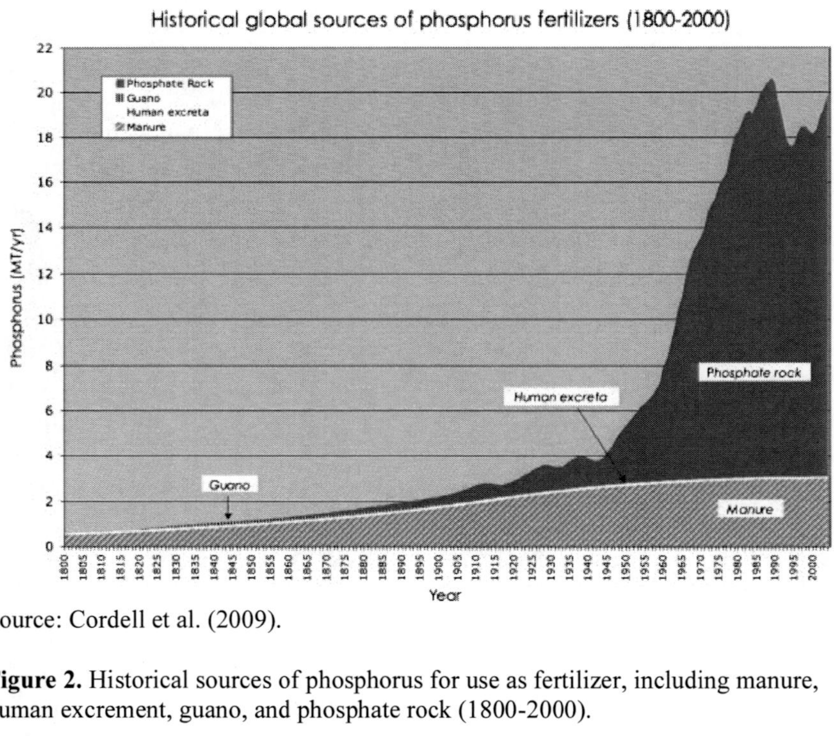

Source: Cordell et al. (2009).

Figure 2. Historical sources of phosphorus for use as fertilizer, including manure, human excrement, guano, and phosphate rock (1800-2000).

Potassium Fertilizers

Potassium (K^+) has several well-known physiological functions and is present in large amounts in plant organisms, corresponding to up to 10% of plant dry weight. The concentration of this nutrient in the cytosol can vary from 100-200 mM, where it is the major inorganic cation. As functions of K in plants, we can cite the control of stomatal opening and membrane electrical potential, regulation of cell osmotic pressure, pH homeostasis, and can neutralize inorganic and organic anions, in addition to macromolecules, such as nitrate, DNA and phospholipids (Said et al., 2018).

In agriculture, K losses occur through the process of leaching, extraction of the nutrient by crops (immobilization), and erosion, a process that depends on the topographic conditions of the land, agricultural practices, and human actions, and that can lead to the eutrophication of rivers and lakes (Chérel et al., 2014). The losses of this nutrient are intensified mainly by the characteristic of potassium fertilizer, which presents high solubility. Thus, the losses are intensified in rainy periods and in areas with a predominance of sand fraction.

K fertilizers are derived from various minerals, such as carnallite and sylvite, which, despite their abundance, are unevenly distributed around the world. Belarus, Canada, and Russia have the world's largest potash deposits, with a concentration of about 60% worldwide, given the quality of these reserves in these countries. Countries such as the United States, Brazil, Germany, and China also have significant deposits of the nutrient.

The efficiency of potash fertilizers depends on fertilizer management. Soils with low levels of the nutrient require successive applications at moderate doses, to minimize losses of the element. The physical quality of the soil, in a way that allows better aeration and less leaching of K, promotes better use of the nutrient by the plants, due to the longer time available for absorption. It is also important to maintain the balance between the potassium fractions in the soil. High applications can disturb the balance between the soil phases and make the nutrient unavailable, by promoting retention with other soil loads. The balance is positive when we consider the production of potassium fertilizers and their global demand. In the year 2017, the estimated value was more than 4100 thousand tons. However, as mentioned before, the global distribution of this element is disuniform, and there are regions where this nutrient is deficient in the soil (Table 2) (Ciceri et al., 2017).

Table 2. Regions with negative potassium (K_2O) balance (Food and Agriculture Organization of the United Nations

Region	Balance 2017 in thousand tons K2O
Africa	−808
Latin America and Caribbean	−5.892
South Asia	−3.552
East Asia	−8.556
Central Europe	−754
Oceania	−387

Source: FAO (2017).

Most mined K deposits originate from ancient oceans. In addition to potassiom chloride, these deposits contain calcium and magnesium salts,, which crystallized as the water evaporated from the ancient oceans. Nowadays, a large part of these reserves lie below the Earth's surface. To give an idea of the size of the exploitation of these reserves, in 2012 around 80% of the world's production of potassium fertilizer came from these underground mines. The other 20% came from natural brine, from water resources with high concentrations of K, through solar evaporation.

The mined ore is milled until the potassium chloride is separated from the sodium chloride. The coarser particles are separated by froth flotation, and then an amine reagent is added, to coat the potassium particles to make them hydrophobic. The addition of the reagent causes the particles to adhere to the air bubbles created and are floated. The sodium chloride particles are deposited at the bottom. The foams formed contain potassium chloride and are removed from the surface, dehydrated, and dried. In this way, the final product is with its concentrated formulation, around 95% potassium chloride, which corresponds to more than 60% K_2O (Ciceri et al., 2017).

The production of mineral fertilizers generally has several negative environmental impacts. In addition to the known problems of mineral extraction, the deforestation of large areas is likely, causing serious damage to the local ecosystem, such as the generation of liquid effluents and solid residues. The processing of the rocks to obtain fertilizers leads to the formation of solid residue, usually contained by dams, which results in the probable contamination of groundwater, especially the water table, as well as the soil and air. Improper disposal of liquid residues causes the eutrophication of rivers and lakes. Environmental legislation monitors the processes involved in this production chain in order to minimize potential problems. However, monitoring andenvironmental laws vary from country to country and even different within the same country.

In addition, to the residues generated by the fertilizer production chain, several industries generate residues. It is not only an environmental but also a social task to try to find ecologically correct alternatives to offer a better destination for these residues.

Alternative Sources of Fertilizers

The use mineral and synthetic fertilizers continues to grow worldwide. Intensive production in cropping systems and the need to increase productivity per area require fast-response inputs to continue producing food. In recent years, the concern for soil health has been one of the most prominent issues in the agricultural world. However, farming systems do not always aim to maintain and improve all soil properties (e.g., chemical, physical, and biological). Soil organic mater (SOM) both its content and quality, and the microbiological community can be affected by soil management, such as the use of these mineral fertilizers, resulting in adverse effects on soil chemical and physical properties (Abebe et al., 2022).

To ensure increased productivity we must take into account the soil, in its complexity and heterogeneity. One of the tools that we can use to improve soil quality is to make use of fertilizers that contain organic matrix, in its total composition, or partially, combined with mineral fertilizers. This union is a tool that, besides collaborating with the quality of the final product, is an environmentally correct way of offering ecologically correct disposal for residues of various natures, that are sometimes disposed of incorrectly in the environment.

Examples of these fertilizers include organic fertilizers, which contain only the organic portion in their composition, and organomineral fertilizers, which contain a combination of the organic and mineral fractions. Both have several advantages for agriculture. In addition, the use of microorganisms in agriculture has increased, and they also have their contribution to soil fertility and can be used in the availability of nutrients for the soil solution. The technology to produce these fertilizers is critical to achieving increasingly sustainable cropping systems, with the maximum plant utilization of the fertilizers applied.

Organic Fertilizers

Organic matter (OM) is a major contributor to the sustainability of agricultural systems and influences several important soil attributes (Thomas et al., 2019; Neal et al., 2020; Cotrufo and Lavallee, 2022). The OM rate depends on the balance of OM outputs in the soil and its inputs. OM outputs occur mainly through erosion and decomposition. The main forms of inputs arethe deposition of plant residues, mulching, or the addition of organic fertilizers (OFs) (Cotrufo and Lavallee, 2022; Jakab et al., 2023). OFs are materials of organic nature, of plant, animal, and industrial origin, which, when applied to the soil at the appropriate time, rate, and manner, directly and indirectly improvethe chemical, physical, and biological properties of the soil. They can be solid or liquid and can be applied in different ways (e.g., leaf, soil, seeds, fertirrigation).

Organic fertilization is a technique used in agriculture since ancient times to supply the nutritional needs of plants (van Raij, 2011). Although mineral fertilization is still the most used, the use of OFs has increased gradually in recent years, mainly in the face of the growing demand for fertilizers, high costs of mineral fertilizers, and environmental problems (e.g., water pollution, acidification, greenhouse gas emissions, soil compaction) (Weifeng et al., 2022).

The advantage of OFs is the recycling of residues from a variety of sources. When it comes to planting sources, the fertilizer can compost leaves, post-harvest residues, stems, pods, and fruits. Another promising organic fertilization technique is green manure, whose purpose is to provide nutrients to the soil mainly nitrogen, through the cultivation of different plants (e.g., mucunas, crotalaria, turnips, oats, guandu, corn, leucena, sorghum) (Adekiya et al., 2019), given the benefits they can bring when incorporated into the soil (Khan et al., 2015). Animal sources can be manure, poultry litter, slaughterhouse residue, pies, etc. (Mateo-Marín et al., 2020; Basso et al., 2021). Residues from industries, urban residue, and sewage sludge can also serve as raw materials (Alves et al., 2021). Due to the different materials and ways of production of these fertilizers, the Brazilian legislation, through decree 4.954, of 14.01.2004, classifies these products into the following categories:

1. Simple organic fertilizer: These are natural products originating from a single source of vegetable or animal origin, for example, animal

manure, sawdust, vinasse, coffee husk, ashes, castor bean cake, bone meal, etc.

2. Mixed organic fertilizer: mix of simple fertilizers, such as castor oil plant cake (high levels of N) mixed with ash (high levels of K);
3. Compound organic fertilizer: product obtained by a physical, chemical, physical-chemical, or biochemical process, natural or controlled, with a mixture of animal, vegetable, industrial, urban, or rural residues, and may be enriched with agents capable of improving its physical, chemical, or biological characteristics. An example of organic compound fertilizer is the Bokashi;
4. Organomineral fertilizers: products resulting from the mixture or combination of mineral and organic fertilizers. An example of organomineral is the mixture of simple superphosphate with animal manure, which is widely used to reduce ammonia losses from manure by volatilization (BRASIL, 2004).

Unlike mineral fertilizers, which make the nutrients available to the plants soon after application, fertilizers withan organic matrix must undergo a mineralization process to convert the organic substances into inorganic ones so that the nutrients can be released to the plants. It is important to plan the timing of the application of OFs, since the gradual release of nutrients depends on the origin of the organic materials and the rate of mineralization, which results in different times for the release of nutrients. This resistance to the decomposition of organic compounds depends on the composition of the material as a whole. Proteins, starches, and sugars decompose faster, followed by cellulose and hemicellulose, compared to waxes, fats, and lignins which decompose more slowly (Liu et al., 2022). To facilitate the mineralization of organic matter, composting is necessary, which promotes the decomposition of organic substances by aerobic heterotrophic organisms in a long thermophilic phase (55ºC to 70ºC), favored by assembly of piles or beds (Chen et al., 2023). The purpose of this technique is to obtain, in the shortest time, a stable material, rich in nutrients and OM, more humified, with properties and characteristics different from the material that gave rise to it (Freire et al., 2013), besides reducing the risks of phytopathogenic microorganisms present in organic materials, that do not resist the heating during the composting period and the high biological activity.

Soil Reactions of Organic Fertilizers

The practice of organic fertilization affects the chemical, physical and biological properties of the soil. The effects on the chemical properties are related to changes in pH, increase in soil buffering capacity, formation of phosphohumates and chelates, that maintain cations in exchangeable forms, maintenance of nitrogen reserves, and increase in cation exchange capacity (CEC). Thus, the addition of OM attracts ions that are dissolved in the soil solution, and acts as a chelator, preventing nutrient losses and providing linear nutrient use by plants (Wu et al., 2020).

The decomposition of the organic material releases several elements, mainly N, P, and sulfur (S) (depending on the organic source used). Besides, the addition of nutrients, OFs reduce P fixation, due to the organic colloids forming immobile complexes with aluminum $Al(OH)_2$ and iron $Fe(OH)_2$, which increases the availability of P by decreasing the fixation of this element (Jin et al., 2023; Zhang et al., 2023). The application of OF has also been recommended as a way to control certain toxic elements found at high rates in the soil; e.g., aluminum (Al), iron (Fe), manganese (Mn), by enabling greater fixation, which controls toxicity (Qian et al., 2013; Kong and Lu, 2022; Saleem et al., 2022).

In soil physical properties organic fertilization reduces thermal oscillations due to the ability of OM to store water and be a low conductor of heat. OM can retain about 4-6 times more water than its own weight, reducing erosion. It changes the consistency of the soil due to the high water absorption by the organic material, which prevents water films around the soil particles and reduces plasticity. The infiltration capacity is also affected due to the increased aggregation and structuring of the soils through the cementing materials, preventing the formation of impermeable crusts (Rós et al., 2020).

It acts directly on the biological properties of the soil, favoring the solubilization of mineral compounds and the health status of underground organs, favoring seed germination and radial respiration, and regulating microbial activity. Theorganic component is a source of nutrients and energy for heterotrophic organisms, keeping the soil in a constant dynamic (Abbott et al., 2018).

Recommendation of Organic Fertilizers

Organic fertilizers have low nutrient content when compared with mineral fertilizers. They can also have unbalanced nutrient contents, when trying to provide the amount of the limiting nutrient in the crop, as a result of increasing the content of other nutrients present in the fertilizer (Zonta et al., 2021).

Technical guidance is important for the use and management of organic-based agricultural inputs. Before fertilization on acidic soils, liming should be carried out according to the soil analysis recommendations.

The application of OFs should precede planting or sowing for better utilization of the nutrients and should be performed on days with mild or cloudy temperatures, or before the rain, to reduce N losses by ammonia volatilization. The application should be associated with conservationist techniques (i.e., crop rotation, no-till, minimal tillage) to reduce nutrient losses in the soil, whether by leaching, erosion, volatilization, or other means.

For the recommendation of OF doses, it is necessary to know the chemical composition of the fertilizer and the soil, the type of soil, the crop to be grown, the application technique, and the relative costs. Therefore, the recommendation can be made based on the recommended doses for each crop and region and according to the crop recommendation, based on the chemical analysis of the soil and the amount of nutrients to be added to the system.

Organomineral Fertilizers

Organomineral fertilizers (OMFs) are products resulting from combinations of organic sources (e.g., animal manure, plant residues, urban wastes) with mineral fertilizers (Smith et al., 2020). To be considered an OMF it must have a minimum organic carbon concentration of 8% for solid product and 3% for liquid product; a maximum moisture of 20% for solid product; a cation exchange capacity (CEC) of at least 80 $mmol_c\ kg^{-1}$ for solid product; Primary macronutrients alone, i.e., nitrogen (N), phosphorus (P), potassium (K) or in mixtures (NP, NK, PK or NPK) minimum of 10% for solid products, and may be added secondary macronutrients, i.e., calcium (Ca), magnesium (Mg) and sulfur (S), or micronutrients of at least 1% and for fluid products, the primary macronutrients alone or in the mixture must have a minimum of 3% (BRASIL, 2020).

Organic fertilizers have low nutrient concentrations, so mineral fertilizers can be used as a supplement to meet the nutrient demands of plants, providing the nutrients in a synchronous release throughout the crop cycle (Li et al., 2017). The organic matrix of the fertilizer acts on the soil attributes, maintaining adequate nutrient contents, minimizing losses by fixation and precipitation of P, volatilization, and leaching of N and K losses by leaching are reduced by presenting gradual solubilization (Redding et al., 2016; Agegnehu et al., 2016) When associated with mineral fertilizers there is an increase in the contents of these elements in the system, which provides balance and utilization of nutrients. Studies comparing OMF with mineral fertilizers conducted in soybean and corn crops found increases in dry mass production when using OMF (Domingos et al., 2018; Bastos et al., 2021). This effect is due to the gradual solubilization of OMF during plant growth, thus increasing agronomic efficiency (Crusciol et al., 2020).

The OMF market has been gaining space in agriculture since it allows the recycling of residue from various sectors of the production chain. Besides allowing the production of specific formulations for each agricultural crop, with the increase of uniformity and concentration of nutrients, which influences smaller applications and consequently reduction of costs in the field, with economic and environmental benefits (Crusciol et al., 2020). OMFs provide anaverage 20% reduction in mineral fertilizer use by providing residual effects, increasing microbial activity, making a higher proportion of nutrients available in the soil, and minimizing losses to the environment. OMFs are classified according to the raw materials used in their production into:

- *Class A:* It uses in the production of raw materials generated in agricultural, agro-industrial, industrial, and commercial activities, free from sanitary dumping or contaminants, resulting in a product that is safe for agriculture;
- *Class B:* Uses any amount of organic raw material generated in urban, industrial, and agro-industrial activities, including sludge generated in sewage treatment and sludge from residual water treatment systems, and all whose use is authorized by the Environmental Agency (BRASIL, 2020).

The application of OMF can be done in the same way as mineral fertilizers, so it can be applied when preparing the soil, sowing or planting and in cover. The OMFs whose organic matrix is classified as class A of animal origin have their use restricted in pasture areas, it will only be allowed if the fertilizer is incorporated into the soil and the entry of animals is only allowed 40 days after the incorporation of the fertilizer. The use of class B fertilizers is forbidden in pasture areas and in vegetables or any crop whose edible part comes into direct contact with the soil (BRASIL, 2020).

As advantages of OMFs, it is expected that, besides the mineral nutrient contribution, they provide all the benefits of organic fertilizers (as detailed in the section "Organic Fertilizers - Reactions in the Soil"): increase in soil CEC, neutralization of toxic elements, reduction in losses of K, N, and greater efficiency in the availability of P, enhancement of microbial activity, increase in water retention, greater contribution of organic carbon in the system, besides providing safe disposal of residues and reduction of greenhouse gas emissions.

Potential Contamination of Alternative Fertilizers

The search for nutrient recycling throughecosystem processes in agro-systems, or technology through fertilizers based on organic residues, is justified by the high demand and dependence on chemical fertilizers and the predictable global energy crises, which generally increase production costs in agriculture and threaten food security. In addition, the use of alternative fertilizers in agriculture is increasing due to the growing interest in more sustainable production systems and methods, such as organic farming, and the growing demand for organic food (FAO, 2021).

Farmers apply chemical fertilizers and/or animal manures and organic composts to fields to provide the nutrients necessary for proper crop growth and economically viable food production. However, when the applied nutrients are not fully utilized by the growing plants, they can be lost from agricultural fields and negatively impact soil, water, and air quality (USEPA, 2022). Thus, even if among the proposals for the use of alternative fertilizers is the reduction of nutrient losses due to their lower solubility compared to conventional fertilizers, if poorly used, they can pose risks to the environment and human health, due to the nature of the product and its mismanagement.

When it rains, nutrient runoff from agricultural fields reaches waterways, when the snow melts in certain regions, and can also leach into the soil profile and, over time, into groundwater. Depending on the concentrations, these elements can lead to eutrophication of water bodies and can cause hypoxia - also known as “dead zones” - causing fish kills and reduced aquatic life. At high concentrations, nutrients can cause harmful algal blooms in freshwater systems, which in addition to disrupting wildlife, can also produce toxins harmful to humans (Khan et al., 2018; USEPA, 2022).

Table 3. Classification of residue according to its origin

Type of Residue	Description	Precautions
Residues from agricultural activity and	Residues of strictly agricultural origin, such as crop residues and manure.	This type of residue is unlikely to contain a high concentration of contaminants, and depending on the composition and the location where the residue is generated, the best option may be to recycle it in local agriculture. Some precautions should be taken, e.g., pig manure may contain high amounts of Cu and Zn due to animal handling, thus its continuous application to the soil can cause nutrient imbalance for crops. The calibration of the dose of residue to be used is fundamental to avoid environmental liabilities (such as the contamination of water sources due to nitrate leaching).
Residues from industrial activity that uses agricultural raw material and in which no element foreign to the agricultural activity is introduced in the industrial process	E.g.: Residues from the sugar and alcohol industry (filter cake, soot and vinasse). This activity uses raw material of agricultural origin (sugarcane) and its industrial process, both for the production of sugar and alcohol, uses only materials usual to the agricultural activity (lime, sulfur and polyphosphates).	
Residues from industrial activity using agricultural raw material into which foreign elements are introduced	Residues generated in food industries generally fall into this class because, in most cases, foreign products are added, such as preservatives, flavorings and others. E.g.: residues from industries that process oils and fats commonly contain nickel, used to catalyze reactions inherent to the process.	Besides the substances added to the food product, generally, cleaning and maintenance products of equipment used in the process end up making up the final residue, and must be considered. E. g.: In the pulp and paper industry, NaOH has the function of separating the cellulose fibers, the sodium cation is not toxic, but if added in large quantities to the soil can cause the dispersion of clays, compromising its quality. Therefore, the dose of residue to be applied is defined from information regarding the total and exchangeable sodium content in the residue.
Residues from industrial activity that does not use agricultural raw material	Residues that generally contain potentially toxic organic and/or inorganic substances in industrialization processes whose final product is not intended for human and/or animal food. As an example, we can cite steel mill dust and electroplating sludge.	For its correct disposal, the packages must be labeled and an inventory of the composition and quantity of the residue must be drawn up, and, according to the legislation in force, the classification of its degree of danger must be indicated. The main disposal methods for this type of residue are Coprocessing - the residues are transformed into fuel for industrial furnaces, without the generation of pollutant gases; Incineration: employed for residues of higher degrees of danger; Landfill - in appropriate landfills, following the norms of each place; Beneficiation of residues - the technical destruction of the contaminable items is carried out.

Type of Residue	Description	Precautions
Urban residues	Residues from urban activities, such as garbage and sewage. These residues, properly separated (selective collection) and prepared, can be composted and transformed into compost, with the possibility of being used in agriculture as organic fertilizer. Among the urban residues are also hospital residues.	The main concern about the use of residue compost in agriculture is the use of residue from the non-selective collection, generating problems of soil contamination with heavy metals, persistent organic pollutants, pathogens, and inert materials (glass, plastics, and others). Sewage sludge can also present such contaminating elements, among others. Hospital residue, due to the risk of contamination, must meet specific legislation, and its incineration has been required, followed by the disposal of ashes in landfills.

Source: Adapted from Pires and Mattiazzo (2008).

Organic fertilizers commonly used in agriculture include animal manure, compost, septic sludge, sewage sludge, municipal biosolids, and food processing residues (Khan et al., 2018). Depending on the source used (more details in Table 3), the risk of contamination depends largely on animal/human health, whichmainly determines microbial concentration, and partially on the presence of antibiotics and chemical components in the organic residue (household or animal sludge), that can be transported to agricultural areas and fertilized crops. Heavy metals can also be applied depending on the source of the residue (agricultural, industrial) (Adegoke et al., 2016; Fijalkowski et al., 2017; Solgi et al., 2018).

In the case of sewage sludge, a material that has been widely studied for use as a matrix in organomineral fertilizers, or for direct application to the soil as compost (Prates et al., 2020; Hawrot-Paw et al., 2022), it is obtained from dewatered and treated municipal wastewater and therefore has high concentrations of hazardous compounds (Halecki et al. 2016), being a mixture of human excreta and other industrial effluents containing metals, organic residues, pharmaceuticals, and pathogens (Östman et al., 2014). However, industrial wastewater and surface soil runoff, however, are the main sources of heavy metals that enrich sewage sludge (Fijalkowski et al., 2017).

In fact, most of the sewage sludge produced in countries with transitional and developing economies does not meet the standards required for direct use in agriculture, either due to high levels of heavy metals resulting from industrial disposal combined with domestic sewage disposal, or because of the high concentrations of pathogens resulting from the poor hygiene and health conditions of the population. However, the regulations and inspections are more stringent in countries with developed economies (USEPA, 1999).

Thus, for the sludge to be applied as soil fertilizer it needs to be transformed into biosolids (USEPA, 2022a), a product that has undergone stabilization with post-treatment aimed at reducing pathogens to achieve the parameters recommended by local agricultural and environmental legislation; thus diluting and balancing the contents of the chemical elements present in its composition, so that they do not exceed the critical limit of soil contamination, and toxicity to plants and organisms. In addition, sludge post-treatment also eliminates undesirable characteristics such as odor and putrefaction, which are also important properties that create resistance to its use in agriculture (Bożym and Siemiątkowski, 2018).

According to USEPA (2022), the following are some of the ways that farmers can reduce their environmental liability by reducing nutrient losses from their operations, including, but not limited to:

- *Adoption of Nutrient Management Techniques:* Applying nutrients (fertilizer and manure) in the right amount, at the right time, with the right method, and in the right place, according to IPNI (2019).
- *Conservation Drainage Practices:* Subsurface tile drainage is an important practice for managing water movement through many soils. Drainage water can carry soluble forms of nitrogen and phosphorus, so strategies are needed to reduce nutrient loads while maintaining adequate drainage for agricultural production.
- *Ensure year-round soil cover:* Sow cover crops or perennial species to avoid bare soil periods in agricultural fields, when the soil (and the nutrients it contains) are most susceptible to erosion and loss to waterways.
- *Planting Field Buffers:* Planting trees, shrubs, and grasses along the edges of fields; this is especially important for fields that border water bodies. Planted buffers can help prevent nutrient loss from fields by absorbing or filtering nutrients before they reach a body of water.
- *Implementation of conservation tillage:* Reduce the frequency and intensity with which fields are plowed. This can reduce soil erosion, runoff, and compaction, and therefore the chance of nutrients reaching waterways through runoff.
- *Managing livestock access to streams:* Farmers and ranchers can install fencing along streams, rivers, and lakes to block livestock access to help restore stream banks and prevent excess nutrients from entering the water.

The increased demands of a growing human population are contributing in different ways to serious environmental problems almost everywhere in the world. Therefore, sustainable consumption patterns that meet human needs, improve the quality of life, and reduce the emission of polluting elements into the environment are urgently needed (Wang et al., 2019). It is currently believed that the benefits of employing organic fertilizers of human and animal origin in fields far outweigh the potential negative impacts (Adegoke et al., 2016). However, the adoption of integrated practices, such as those mentioned above is essential for the development of agriculture along the lines of sustainability.

Legislation on the Manufacture and Commercialization of Alternative Fertilizers

The main differences in hazard identification and characterization regarding the use of these residues as fertilizers in agriculture are locally specific, partly driven by the organic source used (more details in Table 3) and partly reflecting regional and socioeconomic situations (Adegoke et al., 2016). This issue is also related to treatment and application barriers, with regulations and enforcement being more stringent in developed regions and economies (USEPA, 1999).

Thus, as shown in this chapter, the global scientific community agrees that fertilizer related soil, water, and air pollution can be minimized by adopting various control measures; however, public awareness and the establishment of legislation are necessary resources to ensure the effective dissemination of standards for the development of sustainable agriculture.

Regarding the use of organic residues as fertilizers in agriculture, this practice is widespread in European countries, where the continued use of these products over time has triggered significant environmental liabilities, which have been managed under strict regulations in some of the countries. This group of countries presents different policies for the use of residue in agriculture, some stricter (such as Denmark, the Netherlands, and France), others looser (such as Italy, Portugal, and the United Kingdom) (EUR, 1986), especially regarding the use of biosolids as a soil amendments/fertilizers, which represents the most used disposal option in this region (Kelessidis and Stasinakis, 2012).

In this perspective of the utilization of organic residues, attention should be given to biosolids, of which, although here have already been reported some of the main problems related to its disposal in agricultural soils, there is no consensus in the scientific community about the adverse effects caused by application to the soil, and because it is widely spread as a fertilizer in Europe, this fact has led to restrictions on its use based on its quality (Directive 86/278/EEC) (EUR, 1986 and its updated versions EUR 2009; 2018), setting limit values to protect human health through the application of biosolids with good qualities. Based on this directive (Directive 86/278/EEC), each member state of the European community issued its regulations that in some cases set stricter limit values and determined further restrictions, such as limit values for pathogens and organic micropollutants.

Although the specific conditions of each region require standards that are adapted to their differences, the adoption of foreign legislation, from developed countries, can serve as a model for developing countries, as in the case of Brazil where, until mid-2006, the legislation for the application of sewage sludge on agricultural soils was in accordance with the American legislation CFR-40, Part 503 (USEPA, 2003). Since that year, however, the country has developed its own legislation (CONAMA Resolution No 375) (CONAMA, 2006), which is more restrictive than the CFR-40. This can be attributed, in part, to the sanitary conditions of the population and the manner in which municipal and industrial wastewater is disposed of, which contributes to a higher contaminant load in the processing residues.

Therefore, the development of a broad evidence base and its application in a risk management perspective, and framework, will further enhance the benefits of organic fertilizer use in agriculture, and neutralize the negative impacts (Adegoke et al., 2016). This may be possible through regional research aimed at developing fertilizers with minimal contaminants, investigating optimal doses for a given crop, with minimal losses to the environment (Khan et al., 2018).

However, even if different risks and their levels can be identified, epidemiological evidence is still scarce for different types of organic fertilizers and especially whether we should value this transmission route over others. This is also related to different global regions and socioeconomic conditions, i.e., results from specific studies in the United States, the European Union or Australia, for example, cannot be directly extrapolated to conditions and situations on other continents and vice versa (Adegoke et al., 2016).

Bio-Input: An Overview

With the huge and expressive growth of the so-called agribusiness in the world, especially after the 2000s, an economic phenomenon has developed masterfully developed in producing countries like Brazil, with a high import of non-renewable agricultural supplies such as fertilizers and pesticides and the export of primary products. However, the dependence of the agricultural market on foreign policies has increased, the volatility of the dollar, war crises, and strict controls on export exploitation controls by the government and local companies to influence the law of supply and demand. These uncertainties have led to strong demand for renewable resources in various industrial sectors. To this end, companies have formed strong partnerships with research institutes and universities in the agricultural and food sectors. Since there is a need to produce more and more agricultural products and by-products to sustain a hybrid matrix of food contingents.

Some imported fertilizers such as phosphates are required in greater demand in regions where soils are poor and weathered as mentioned earlier. Mosttropical regionshave highly weathered soils with low concentrations of essential elements for plants such as Ca, Mg, P, K, S, and others. In addition, these soils have poor chemical characteristics for the cultivation of most crops such as high acidity and concentration of free aluminum.

The study of the different types of soils becomes more than essential to have a real idea of their nutritional, chemical, physical, and biological demand. It is known that all the balanced biogeochemical cycles that take place in the soil depend on microorganisms. They are responsible for the cycling of organic matter and consequently for the release of nutrients naturally available in the soil. Thus, they are directly responsible for leading to higher crop yields (Lisboa et al., 2012).

In this context, the use of alternative products such as bio-inputs, which provide nutrients and at the same time stimulate the soil microbiota, has become in the current economic scenario one of the best, or if not the best, alternatives in these world areas with high demand for mineral fertilizer imports. It is well known that biological inputs (or bio-inputs) are products derived from microorganisms, plant materials, animal residues, and organic or natural materials and used in agricultural cropping systems to control pests and diseases and/or to improve soil fertility and plant nutrient availability. In terms of providing nutrients to the soil, biofertilizers are the best example of bio-inputs. Biofertilizers are products composed of active components or animal, plant, or a macrobiotic organic substance, that act to increase the productivity, efficiency, and quality of plants by behaving more harmoniously with the soil system.

One question is what is driving the interest in this type of product (alternative crop inputs), after all, they have been in use for a long time, so why the boost at this time? Certainly, the increase in investment in research focused on this area has a strong relationship, and it is undeniable that the knowledge of biological inputs has increased in recent times. Agrifood companies have recognized the potential of understanding the microbiome and translating this knowledge into products (Olmo et al., 2022). Apparently, they have realized that there may be other very interesting sources of nutrients other than the commonly used and finite ones.

Brazil, for example, is one of the few countries in the world that has successfully applied biofertilizers to soybeans. The country is the largest producer and exporter of the commodity and, currently, 80% of the soybean area uses inoculation of microorganisms to fix nitrogen (Olmo et al., 2022). In Figure 3, we can see a study conducted by Hungria et al. (2013), in which the inoculation of bean seeds of common bean inoculation with *Rhizobium tropici* resulted in an average increase of 8.3% in grain yield when compared to the non-inoculated treatment, with a further increase (14.7%) with the co-inoculation of *Azospirillum brasilense* in-furrow. This data shows a significant increase in bean production with the use of bio-inputs. For the farmer, this increase, in addition to saving on fertilizer, represents a huge additional financial gain when extended to the entire planted area.

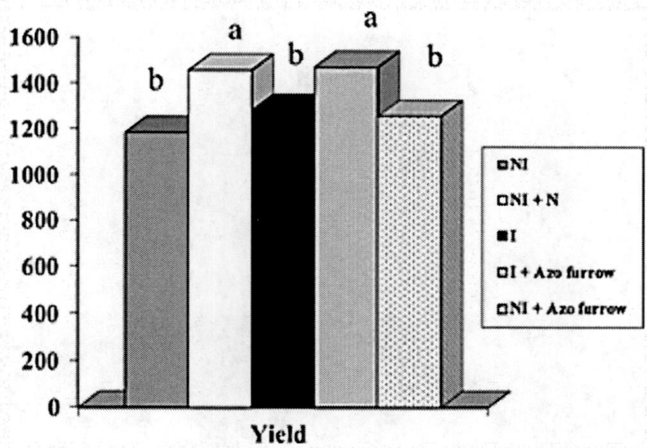

Source: Hungria (2013).

Figure 3. Seed Inoculation effects on common bean grain yield (kg ha^{-1}) with Rhizobium and in-furrow inoculation with *Azospirillum*. Treatments are **NI** (non-inoculated); **NI + N** (non-inoculated + 20 kg N ha^{-1}, split-applied as urea at sowing and 60 kg of N split-applied at 25 DAE); **I** Rhizo, seed inoculation with Rhizobium, 1.2×10^6 cells $seed^{-1}$; **Azo**, inoculation with *Azospirillum* in the furrow, at a concentration of 2.5×10^5 cells seed. Different letters indicate statistical differences ($p \leq 0.05$, Duncan's test).

Co-inoculation has also shown great potential for agricultural benefit. This technique consists of using more than one species of microorganism together in inoculation. It has been the object of interest the interaction between symbiotic bacteria in the case of *Bradyrhizobium* and diazotrophic bacteria belonging to the genus *Azospirillum*.

A large study on the co-inoculation of soybean with *Azospirillum brasilense* and *Bradyrhizobium spp*, was developed by Barbosa et al. (2021), in which data from field experiments were conducted between 2009 and 2020. These data were collected to evaluate which environmental and agronomic conditions affect the response of soybean to bacterial co-inoculation in which we can see increases in grain yield, nodulation, and root growth (Figure 4). The main numerical results of the co-inoculation of soybean were increased root mass in 11%, nodule number in 5.4%, nodule mass in 10.6%, N concentration in shoots in 2.8%, grain yield and N concentration in grains in 3.2% and 3.6%, respectively compared to the conventional single inoculation with *Bradyrhizobium* spp.

Source: Barbosa (2021).

Figure 4. Overview of soybean response to co-inoculation with *Azospirillum brasilense* and *Bradyrhizobium* spp. in Brazil.

An increase in productivity of almost 3% in productivity for beans is very significant. We are in an era in which agricultural production must be increasingly optimized, to produce more without the need to open new areas and with the least use of chemicals possible. This is going to meet with several aims of different Sustainable Development Goals (SDGs) for the fulfillment of the UN 2030 Agenda. Every type of management that leads to greater sustainability of the planet combined with higher food production will be increasingly demanded.

Practical evidence and research results show that working with bio-inputs is very promising: a survey by the Brazilian Agricultural Research Corporation (EMBRAPA) shows that about 95% of farmers believe in the growth of the market of biological products. This confidence is mainly credited due to its contribution to sustainability, productivity, and the economy.

A recent study published by Spark Strategic Intelligence (SSI) points out that in 2020/2021 crop year, the bio-inputs market in Brazil recorded a business volume of R$1.7 billion - a 37% increase over the volume traded in the previous cycle. The market for biological solutions grew by 40% per year between the 2018/2019 and 2020/2021 crops years (SSI, 2021). Specifically, the bioinoculants were responsible for a market turnover of R$ 393 million, equivalent to 23% of the biological market.

In general, when a product receives such a consistent adhesion, the justification is not limited to numbers, on the contrary, they are nothing more than the consequence of a mix of market needs to be satisfied. For those who have already started to work with bio-inputs some time ago, it is possible to identify specific advantages, since when choosing this type of product is chosen, practically customized compositions can be created for specific purposes, but even if we look at biologics in general, we can identify several advantages that serve farmers in general, and they range from safety to economic issues (Machado et al., 2012).

The economic advantages of working as bio-inputs are drawn from various elements of the cost composition of a crop, lower production costs, the financial advantage of bio-inputs appears even before the product reaches the shelves, because its development is much cheaper than that of a chemical, which is reflected in the final price, ie there is a reduction in costs in national and international currencies. As already mentioned, several fertilizers and chemical pesticides used in many agricultural countries are imported, varying according to the exchange rate, especially the dollar, in the case of bio-inputs, in contrast, we have excellent options with national production. Thus, even without giving up the chemicals, the farmers can significantly reduce the consumption of imported chemical inputs in an associated application program. When working with bio-inputs to solve one problem, a positive reflex is identified which a product for biocontrol of a pathogen, for example, can improve the nutrition of the plant.

To illustrate what happens in the use of inoculants, the bio-input use composed of bacteria to help in nitrogen fixation often causes additional effects since these bacteria also can produce plant growth hormones. These hormones stimulate an increase in the density of root hairs, secondary roots, and consequently the total root surface, thereby improving nutrient and water uptake. As a result of this chain reaction, the plant benefits through increased production capacity, as well as being more resistant to environmental challenges.

Biological inputs are biodegradable since they are produced with microorganisms, plant materials, and organic or natural materials. This translates into a significant decrease in the environmental risks of agricultural activity. Besides the environmental issue, working with bio-inputs brings advances for the safety of the farmer and the consumer. After all, it is a type of product with very low toxicity, whose residues do not bring harm to the soil, food, or health of those who consume them.

Certainly, working with bio-inputs, for sure, is a way to contribute to the environmental sustainability, because the impact is are much less when natural products are applied to the soil. In this case, we also have the advantage of using products that for their development considered the relationship of the elements involved with the ecosystem. Thus, they strategically use the dynamics of nature to preserve it, so that, for example, the carbon generated in agricultural production is not lost to the atmosphere, but is stored in the soil in the form of stabilized organic matter.

Similarly, several of the residues generated in the crop's production end up being recycled by microorganisms present in the soil. The reduction in water uptake is another consequence that is observed when working with bio-inputs, as they contribute to the structuring of the soil, improving biota health and plant nutrition. The roots develop in a way that contributes to the better use of both, water and soil nutrients, again, a chain effect, however this time it is reflected in the reduction of water needs.

Choosing to apply bio-inputs in the field does not necessarily imply radically abandoning chemical inputs. There are still situations where biological solutions are not the most appropriate, so the use of biologicals can compose an integrated management (Machado et al., 2012). Working with bio-inputs is an approach to soil treatment that prevents factors that can threaten continued production, when a "curative action" is needed, the tools are at hand. The idea is that these emergencies give way to healthy soil and the presence of a balanced microbiome, ideal for sustainable cultivation, both in environmental and economic terms.

Inhibitors, Controlled and Slow-Release Fertilizers

There are many reasons to use slow and controlled-release fertilizers. Throughout the crop cycle, different biotic and abiotic reactions occur in the soil involving the essential elements. Both, nutrients added through fertilization and those already in the soil and those coming from biogeochemical cycling of organic matter have the risk of suffering losses and leaving the zone of uptake by the plants. Among the nutrient losses, the main phenomena are; runoff, volatilization, leaching (physical process); exchange, fixation, precipitation, hydrolyses (chemical); nitrification, denitrification, and immobilization (biological/microbial process) (Trenkel, 2010). Such losses need to be avoided as much as possible since it involves the misuse of a limited resource, and this leads to a decline in productivity and possible environmental problems. Currently, with the significant increase in the price of fertilizers due to the international crisis, shocks in the acquisition of these, and their consequent production, the problematization of mineral fertilizer use is very evident. Thus, better management and use of technologies are increasingly employed to reduce these losses.

There are numerous possible ways to avoid and mitigate the losses of fertilizers applied throughout the crop cycle. An example that can be cited is the utilization of certain substances that are used to chemically or physically preserve the fertilizer. It is known that each nutrient has a dynamization that can lead to its losses.

Nitrogen is undoubtedly the nutrient most taken up by most crops. However, the efficiency of N use by plants is relatively low. Some studies show that on the world average, only 50% of the N fertilizers used in fertilization are effectively utilized by the crops and that the recovery efficiency for cereal production (e.g., sorghum, corn, wheat, barley, oats, rye, and rice) is approximately 33% (Hawkesford, 2011). This N loss is not utilized by living organisms (plants, microorganisms, etc). The risk of this loss depends on the type of nitrogen source you are applying, your soil type, temperature, and management practices.

The most appropriate management for N fertilizers is their incorporation into the soil, thus increasing the protected area around the fertilizer granule, which contributes to the transformation of NH_3 into NH_4^+. A major constraint of incorporation is the requirement of specific equipment, cost, and an operation with low operational yield. Nitrogen loss is a challenge that every farmer faces, whether applying Urea or other sources, regardless of the season. As you begin to prepare your nitrogen fertility plans, it is important to understand how to mitigate each of these losses. As demonstrated in Figure 5, it´s possible to see all forms of N transformations in the soil.

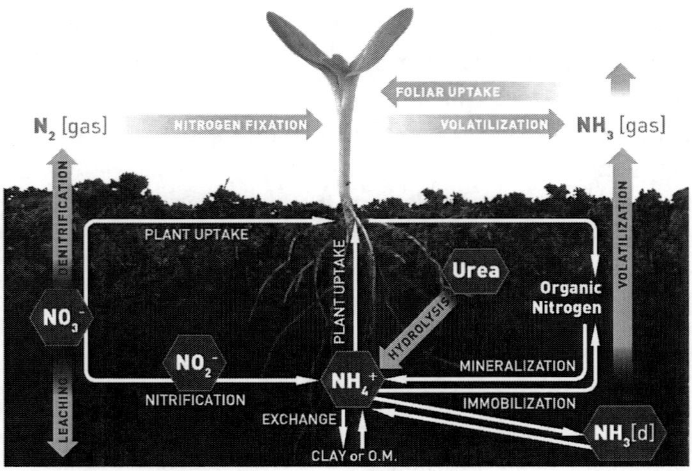

Source: Taurus (2022).

Figure 5. Nitrogen cycle and management strategies.

As already mentioned, certain substances are used to chemically or physically preserve the fertilizer. From this, urea sources with urease inhibitors, which imply a slower release of N from the pellet, have remarkable importance in maximizing N application and utilization by the crops. The most known and widespread urease inhibitor is NBPT (n-N-butyltriamide thiophosphate or N-butyltriamide thiophosphoric acid).

Under flooded soils with low aeration, the conversion of NBPT to NBPTO is slow, decreasing the efficiency of nitrogen application. NBPT decelerates the rate at which urea dissociates, which mitigates the effect of raising the pH around the granule. Even so, the product (NBPT) alone does not solve 100% of the volatilization problem and the ideal is that the fertilizer, even protected, be incorporated via water (rain or irrigation) or mechanically into the soil. To inhibit urease, when NBPT is in the soil, it is transformed into N-n-butyltriamide phosphate (NBPTO), which acts by inactivating the enzyme urease. It can inhibit urea hydrolysis for a period of 3 to 14 days, this period being very dependent on soil and climate conditions. In a study related to the inhibition of urease hydrolysis was found that at the temperature of 25°C in wheat-grown soils, NBPT greatly decreased the rate of urea hydrolysis, but on days 14 and 15, less than 2% of the urea was present in soil (Suter et al., 2011).

Other technologies to inhibit urease and decelerate the rate of N losses exist on the market. An example is the use of elements from the metal groups in the manufacture of fertilizer under some conditions, and these elements can also be used as nutrients by plants. These include copper (Cu), molybdenum (Mo), cobalt (Co), zinc (Zn), manganese (Mn), silver (Ag), lead (Pb), mercury (Hg); these ions react with the sulfhydryl group, blocking the active site of the enzyme (Adotey et al., 2017). Also, nitrification inhibitors, by reducing the action of bacteria. All this preserves the ammoniacal nitrogen (NH_4^+) in the soil which is a form less susceptible to losses. These inhibitors can be used in various sources of N, such as ammonium nitrate, and even in urea itself. Among other compounds, some examples are nitrapyrin, dicyandiamide (DCD), and 3,4-dimethylpyrazole phosphate (DMPP), each inhibitor is an indicator for a condition, as they are extremely affected by the type of soil and its climatic condition.

Slow and controlled release fertilizers are so called when compared to a soluble reference source, including encapsulated, slowly water soluble, and water-insoluble fertilizers. These fertilizers release nutrients in a way that delays their availability for uptake or extends their availability so that the release is in sync with the plant's nutrient requirements. There is no official distinction between slow-release and controlled-release fertilizers. However, slow-release fertilizers have the differential of prolonging for a longer time the availability of the element (Nascimento et al., 2013) This fertilizer has a slower rate of release than the conventional ones, but not a very controlled rate, pattern, and duration of release. Slow-release fertilizers are chemically modified, one of the most widely used is urea formaldehyde, isobutylaldehyde, and crotonaldehyde (Souza et al., 2017).

When comparing [Mg-$AlPO_4$]-HDL with a conventional source KH_2PO_4 in a wheat experiment, it was found that this nano-fertilizer increased soil pH, which makes P available to plants beyond 30 days, and that the release of P in water from [Mg-$AlPO_4$]-HDL was 10-fold slower than from KH_2PO_4, which demonstrates the potential of [Mg-Al-PO_4]-HDL to be used as a slow-release phosphorus source for plants (Bernardo et al., 2017). Another example of slow-release phosphate fertilizer is struvite which is a mineral formed by reacting magnesium, ammonium, and phosphate to produce a crystalline solid, has a slow release rate due to low water solubility, which leads to a reduction in leaching losses and a prolonged residual value for crops. The struvite can be produced from the recovery of phosphorus present in domestic sewage and other sources as pig slurry, with greater agricultural efficiency than conventional fertilizers. Was compared struvite (St) application with triple superphosphate (TSP) in Soybean cultivation, which was found that St composites provided superior biomass compared to a reference of triple superphosphate with ammonium sulfate (AS), with up to 3 and 10-fold higher mass of shoots and roots, respectively (Valle et al., 2022).

Controlled-release fertilizers have a predictable nutrient availability, which means they have a rate pattern and duration of nutrient release, different from slow-release fertilizers, due to the coating that covers the granule. Mostly controlled-release fertilizers available on the market are those coated with sulfur and polymers as thermoplastics (and sometimes ethylene-vinyl acetate and surfactants, etc.) polystyrene, polyesters, polyurethane, fatty acids latex, petroleum-based products (resins), calcium phosphate, gypsum, and wax. Thus, the availability of the nutrient to the environment depends on the thickness and quality of the granule coating (Trenkel, 2010). The availability of the nutrient from these fertilizers occurs through the granule exposure to weathering agents, such as water, temperature, chemical agents, and mechanical impacts, making possible the rupture of the protective film. We can see a schematic mechanism of a controlled-release fertilizer coated by a polymer in Figure 6.

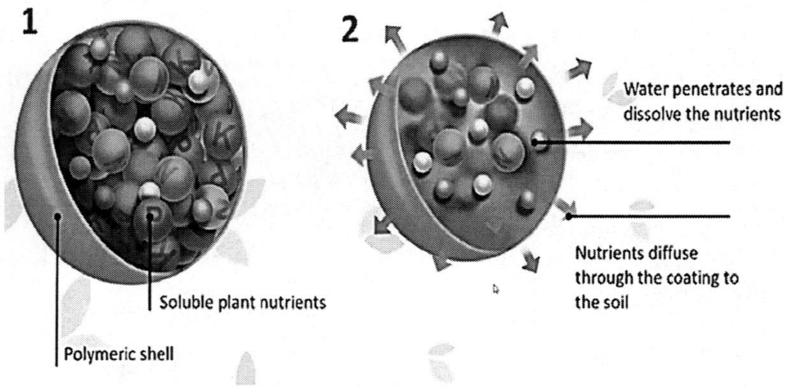

Source: Adapted from Haifa-group (2022).

Figure 6. Mechanism of controlled-release fertilizer coated by a polymer layer.

The coating allows water to diffuse slowly through, wetting the fertilizer which causes granule dissolution over time. This dissolved fertilizer then moves through the coats into the soil via diffusion (Verburg, 2020). In a more detailed way, first, the fertilizer inside the coated pellet is completely free of moisture. The coating allows water to move slowly inside the pellets, which after a period of time dissolves the fertilizer. With the fertilizer dissolved, the nutrient then moves through the coating and becomes available to the crop. Thus, the concentration of fertilizer inside the pellets decreases, and this difference in concentration causes a gradual slowing of the release rate to occur, a representative scheme can be found in Figure 7.

Source: Adapted from Verburg (2022).

Figure 7. Scheme of dry pellets of controlled-release fertilizer.

Another point to emphasize is the high aggregated value and efficiency of these technologies when compared to the common fertilizer, which makes these fertilizers to be used, mainly in crops with high production costs and return on investment such as cotton, sugarcane, vegetables, etc. In recent years, the rapid development of coated fertilizers accounts for more than 95% of controlled-fertilizer (Fu et al., 2018). Concerning these covered fertilizers, as there is a coating around the granule, this treatment can also be used with other nutrients, besides nitrogen. Depending on the production technology, fertilizer prices increase in the following order: conventional < stabilized < blends < slow release < controlled release (Guelfi, 2017).

As the exclusive use of slow-release or controlled-release fertilizers presents a high cost, "blends" has become an interesting alternative to overcome this situation, as they present lower costs (Freitas, 2020). Nitrogen fertilizer blends consist of a physical mixture of different classes of fertilizers. According to Noellsch et al., (2009), most often this mixture is made with conventional fertilizers (with high solubility in water) and controlled-release fertilizers (low solubility). Thus, the conventional fertilizer provides N to meet the initial crop demand and the controlled-release fertilizer for a longer period in further stages.

There are the following different examples of nitrogen fertilizers: *conventional* - Conventional Urea, Ammonium Nitrate, and Ammonium Sulphate + $CaCO_3$; *slow-release* - Urea + formaldehyde; *controlled release* -

Urea + polyurethane and Urea + plastic resin and *blends* (stabilized + controlled release) i.e., - Polyblen Extend® (Wei et al., 2020).

Biochar can be an excellent matrix for creating fertilizers that release nutrients slowly (slow-release fertilizers and controlled-release fertilizers). It is characterized by high specific surface area and porosity, as well as the presence of various functional groups (Marcińczyk and Oleszczuk, 2022). By definition, biochar is a heterogeneous organic matrix, rich in aromatic carbon and mineral elements present in its ashes, produced by pyrolysis of biomass obtained by sustainable criteria, through clean technologies, and under controlled conditions (Maia et al., 2021). Thus, every source of plant biomass can be a source for production and biochar. Three important strategies are strongly recommended within the biochar concept: organic residue recycling, renewable energy generation, and low-carbon agriculture.

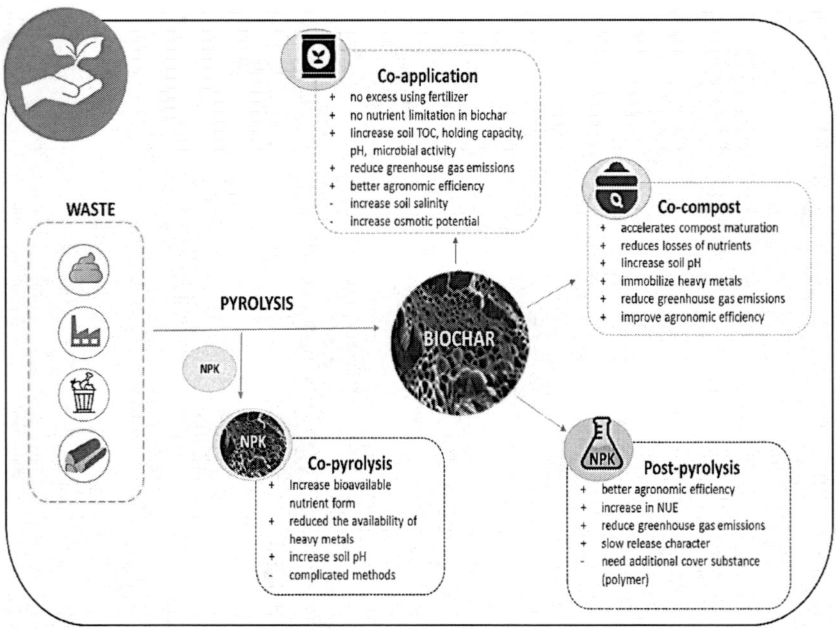

Source: Marcińczyk and Oleszczuk (2022).

Figure 8. Production of nutrient-enriched biochar.

The use of biochar as a source of nutrients is very common among many indigenous peoples, such as indigenous peoples. For a long time, the "Terra preta de Indio" soil has been a source of study of researchers since this soil has Physico-chemical properties different from the local neighboring soils. It is rich in nutrients and has improved ion exchange properties and this improvement probably happens due to the modifier power of biochar (Ye et al., 2020).

Some very recent studies have shown that combining biochar with inorganic fertilizers has been a very interesting partnership, as the plant has become more efficient in utilizing the nutrients. According to Marcińczyk and Oleszczuk (2022), a very viable alternative is to add some mineral source of macronutrient in combination with the compost after the pyrolysis of the organic residues and incorporate, or to enrich it with nutrients through co-pyrolysis (Figure 8).

Conclusion

Given the dependence on fertilizer imports from several countries by large producing and exporting countries, the search for total or at least partial autonomy in the manufacture of these products is necessary to ensure productivity, without being held hostage to exchange rate variations. It is a fact that there are a range of environmental issues that involve in the manufacturing and processing of mineral fertilizers. At the same time, there is a large amount of residue generated by industrial and urban centers. Looking at both sides of the coin, the use of this residue as fertilizer can not only supply the nutritional requirements of plants but also promote the physical, chemical, and biological quality of the soil, while providing an environmentally friendly destination for the residue. Thus, it is crucial that all agricultural sectors mobilize to seek autonomy in the production of fertilizers, thinking not only of not being dependent on the external market but also of the environment, since these mineral fertilizers come from finite sources.

References

Abbott, L. K., L. M. Macdonald, M. T. F. Wong, M. J. Webb, S. N. Jenkins, and M. Farrell. 2018. Potential roles of biological amendments for profitable grain production – A review. *Agriculture, Ecosystems & Environment*, *256*:34–50.

Abebe, T. G., M. R. Tamtam, A. A. Abebe, K. A. Abtemariam, T. G. Shigut, Y. A. Dejen, and E. G. Haile. 2022. Growing Use and Impacts of Chemical Fertilizers and Assessing Alternative Organic Fertilizer Sources in Ethiopia. *Applied and Environmental Soil Science*, 2022:4738416.

Adegoke, A. A., O. O. Awolusi, T. A. Stenström, A. A. Adegoke, O. O. Awolusi, and T. A. Stenström. 2016. Organic Fertilizers: Public Health Intricacies. *Organic Fertilizers - From Basic Concepts to Applied Outcomes*. https://doi.org/10.5772/64195

Adekiya, A. O., T. M. Agbede, C. M. Aboyeji, O. Dunsin, and J. O. Ugbe. 2019. Green manures and NPK fertilizer effects on soil properties, growth, yield, mineral and vitamin C composition of okra (Abelmoschus esculentus (L.) Moench). *Journal of the Saudi Society of Agricultural Sciences*, *18*(2):218–223.

Adotey, N., M. Kongchum, J. Li, G. B. Whitehurst, E. Sucre, and D. L. Harrell. 2017. Ammonia Volatilization of Zinc Sulfate-Coated and NBPT-Treated Urea Fertilizers. *Agronomy Journal*, *109*(6): 2918–2926.

Agegnehu, G., P. N. Nelson, and M. I. Bird. 2016. The effects of biochar, compost and their mixture and nitrogen fertilizer on yield and nitrogen use efficiency of barley grown on a Nitisol in the highlands of Ethiopia. *Science of The Total Environment*, *569–570*:869–879.

Alves, B. S. Q., K. P. S. Zelaya, F. Colen, L. A. Frazão, A. Napoli, S. J. Parikh, and L. A. Fernandes. 2021. Effect of sewage sludge and sugarcane bagasse biochar on soil properties and sugar beet production. *Pedosphere*, *31*(4):572–582.

Barbosa, J. Z., M. Hungria, J. V. D. S. Sena, G. Poggere, A. R. Reis, & R. S. Corrêa. 2021. Meta-analysis reveals benefits of co-inoculation of soybean with *Azospirillum brasilense* and *Bradyrhizobium* spp. in Brazil. *Applied Soil Ecology*, 163, 103913.

Basso, C. J., M. A. B. Pinto, R. Gonzatto, S. B. Pujol, and F. M. Souza. 2021. Management of liquid swine manure: impact on mineral nitrogen dynamics and corn yield. *Pesquisa Agropecuária Tropical [Tropical Agricultural Research]*, *50*:1–10.

Bastos, A. V. S., M. B. Teixeira, F. A. L. Soares, E. C. Silva, L. N. S. Santos, and F. H. F. Gomes. 2021. Immediate and Residual Effects of Mineral and Organomineral Nitrogen Sources Associated with Concentrated Vinasse on Maize. *Journal of Soil Science and Plant Nutrition*, *21*(2):1382–1396.

Battye, W., V. P. Aneja, and W. H. Schlesinger. 2017. Is nitrogen the next carbon? *Earth's Future*, 5(9):894–904.

Bernardo, M. P., G. G. F. Guimarães, and C. Ribeiro. 2017. Fertilizante fosfatado de liberação controlada baseado em hidróxidos duplos lamelares. [Controlled-release phosphate fertilizer based on lamellar double hydroxides.] *IX Workshop de Nanotecnologia Aplicada ao Agronegócio*. [*IX Workshop on Nanotechnology Applied to Agribusiness*.] Available at:< https://ainfo.cnptia.embrapa.br/digital/bitstream/item/168031/1/P-Fertilizante-Fosfato-de-Liberacao-Controlada.pdf> Accessed at: November 22, 2022.

Bożym, M., and G. Siemiątkowski. 2018. Characterization of composted sewage sludge during the maturation process: a pilot scale study. *Environmental Science and Pollution Research*, *25*(34):34332–34342.

BRASIL. Decreto nº 4.954, de 14 de janeiro de 2004. [Decree No. 4,954, of January 14, 2004.] Aprova o regulamento da lei n.º 6.894, de 16 de dezembro de 1980, que dispõe sobre a inspeção e fiscalização da produção e do comércio de fertilizantes, corretivos, inoculantes ou biofertilizantes destinados à agricultura, e dá outras providências. [Approves the regulation of Law No. 6,894, of December 16, 1980, which provides for the inspection and supervision of the production and trade of fertilizers, correctives, inoculants or biofertilizers intended for agriculture, and provides other measures.] *Decreto nº 4.954 de 14 de janeiro de 2004*, [*Decree No. 4,954 of January 14, 2004*], Brasília, 2004.

BRASIL. *Instrução Normativa n° 61, de 08 de julho de 2020.* [*Normative Instruction No. 61, of July 8, 2020*] 2020. Estabelece as regras sobre definições, exigências, especificações, garantias, tolerâncias, registro, embalagem e rotulagem dos fertilizantes orgânicos e dos biofertilizantes, destinados à agricultura. [Establishes rules on definitions, requirements, specifications, guarantees, tolerances, registration, packaging and labelling of organic and biofertilizers intended for agriculture.]

Cameron, K. C., H. J. Di, and J. L. Moir. 2013. Nitrogen losses from the soil/plant system: a review. *Annals of Applied Biology*, 162(2):145–173.

Chen, L., Y. Chen, Y. Li, Y. Liu, H. Jiang, H. Li, Y. Yuan, Y. Chen, and B. Zou. 2023. Improving the humification by additives during composting: A review. *Waste Management*, *158*, 93–106. Chérel, I., Lefoulon, C., Boeglin, M., and Sentenac, H. 2014. Molecular mechanisms involved in plant adaptation to low K(+) availability. *Journal of Experimental Botany*, *65*(3):833–848.

Ciceri, D., M. Oliveira, A. Allanore. 2017. Potassium fertilizer via hydrothermal alteration of K-feldspar ore. *Green Chemistry*, 19:5187-5202.

CONAMA (Conselho Nacional Do Meio Ambiente). 2006. *Resolução n° 375, de 29 de agosto de 2006.* [*Resolution No. 375 of August 29, 2006*] https://www.normasbrasil.com.br/norma/resolucao-3752--006_103464.html.

Cordell, D., J. O. Drangert, and S. White. 2009. The story of phosphorus: Global food security and food for thought. *Global Environmental Change*, 19(2):292–305.

Cotrufo, M. F., and J. M. Lavallee. 2022. Soil organic matter formation, persistence, and functioning: A synthesis of current understanding to inform its conservation and regeneration. *Advances in Agronomy*, *172*:1–66.

Crusciol, C. A. C., M. Campos, J. M. Martello, C. J. Alves, C. A. C. Nascimento, J. C. R. Pereira, and H. Cantarella, 2020. Organomineral Fertilizer as Source of P and K for Sugarcane. *Scientific Reports, 10:1*, *10*(1):1–11.

Domingos, C. F. K., M. J. F. Scherrer, A. J. J. Júlio, S. G., André, M. B., Campos, L. A. Mendes, and L. M. Silva. 2018. Desempenho agronômico da soja convencional cultivada com fertilizantes organomineral e mineral. [Agronomic performance of conventional soybean cultivated with organomineral and mineral fertilizers] *Nucleus*, *15*(2):301–309.

Engin Sagirli, F. Z., E. S. Kayali, A. S. Sarac -, L. Shen, Y. Ma, Y. Wang, J. Harding, and X. Tu. 2021. Plasma-enhanced N2 fixation in a dielectric barrier discharge reactor: effect of packing materials. *Plasma Sources Science and Technology*, 30(10):105002.

EUR-Lex. 1986. Council Directive 86/278/EEC of 12 June 1986 on the Protection of the Environment, and in Particular of the Soil, when Sewage Sludge is Used in Agriculture. *Official Journal of the European Communities*, 181:6–12.

EUR-Lex. 2009. Regulation (EC) No 219/2009 of the European Parliament and of the Council of 11 March 2009 adapting a number of instruments subject to the procedure referred to in Article 251 of the Treaty to Council Decision 1999/468/EC with regard to the regulatory pro. *Official Journal of the European Communities* 87:109–154.

EUR-Lex. 2018. Decision (EU) 2018/853 of the European Parliament and of the Council of 30 May 2018 amending Regulation (EU) No 1257/2013 and Directives 94/63/EC and 2009/31/EC of the European Parliament and of the Council and Council Directives 86/278/EEC and 87/217/EEC. *Official Journal of the European Communities* 150:155–161.

FAO (Food and Agriculture Organization of the United Nations). 2017. *World Fertilizer Trends and Outlook to 2020.* Available from: https://www.fao.org/3/i6895e/i6895e.pdf (accessed November 20, 2022).

FAO (Food and Agriculture Organization of the United Nations). 2021. G*lobal assessment of soil pollution: report.* https://doi.org/10.4060/cb4894en (accessed October 21, 2022).

Fijalkowski, K., A. Rorat, A. Grobelak, and M. J. Kacprzak. 2017. The presence of contaminations in sewage sludge - The current situation. *Journal of Environmental Management*, *203*(Pt 3):1126–1136.

Freire, L. R.; F. C. Balieiro, E. Zonta, L. H. C. Anjos, M. G. Pereira, E. Lima, J. G. M. Guerra, M. B. C. Ferreira, M. A. A. Leal, D. V. B. Campos, J. C. Polidoro. 2013. *Manual de calagem e adubação do Estado do Rio de Janeiro*. [*Manual of liming and fertilization of the State of Rio de Janeiro*.] 1. ed. Rio de Janeiro: Embrapa, 1:430.

Freitas, T. 2020. F*ertilizantes de liberação controlada e seus "blends" com ureia tratada com inibidor de urease para o cafeeiro*. [*Controlled-release fertilizers and their "blends" with urea treated with urease inhibitor for coffee*.] Thesis (Doctorate´s degree agronomy). Federal University of Lavras, Lavras, Brazil.

Fu, J., C. Wang, X. Chen, Z. Huang, and D. Chen. 2018. Classification research and types of slow controlled release fertilizers (SRFs) used - a review. *Communications in Soil Science and Plant Analysis*, 49(17):2219–2230.

Geng, C., J. Li, T. Weiske, and H. Schwarz. 2019. Complete cleavage of the N≡N triple bond by Ta2N+ via degenerate ligand exchange at ambient temperature: A perfect catalytic cycle. *Proceedings of the National Academy of Sciences of the United States of America*, *116*(43):21416–21420.

Guelfi, D. 2017. Fertilizantes nitrogenados estabilizados, de liberação lenta ou controlada. [Stabilized, slow-release or controlled nitrogen fertilizers.] *Informações agronômicas*, [*Agronomic information*] 157:1-14. Informações Agronômicas, Belo Horizonte, 157:1-14.

Haifa-group. 2022. *Polymer coating controlled-release technology.* Available at:< https://www.haifa-group.com/articles/multicote-technology> Accessed: November, 21, 2022.

Halecki, W., M. Gasiorek, F. Gambus, and R. Abram. 2016. The potential of hydrated and dehydrated sewage sludge discharges from soil reclamation appliances. *Fresenius Environmental Bulletin,* 25(6):1935-1941.

Hawkesford, M. J. 2011. An Overview of Nutrient Use Efficiency and Strategies for Crop Improvement In *The Molecular and Physiological Basis of Nutrient Use Efficiency in Crops*, ed. Hawkesford, M. J. and Barraclough, 3-19. New Jersey: John Wiley & Sons.

Hawrot-Paw, M., M. Mikiciuk, and A. Koniuszy. 2022. Meller, E. Influence of Organomineral Fertiliser from Sewage Sludge on Soil Microbiome and Physiological Parameters of Maize (*Zea mays L.*). *Agronomy*, 12(5):1114.

Heffer, P., and M. Prud'homme. 2016. *Global nitrogen fertiliser demand and supply: trend, current level and outlook Figure 1: Evolution of global fertilizer consumption by nutrient (Tg N+P_2O_5+K_2O) (IFA, 2016a)*. www.ini2016.com.

Hungria, M., M. A. Nogueira and R. S. Araújo. 2013. Co-inoculation of soybeans and common beans with rhizobia and azospirilla: strategies to improve sustainability. *Biol Fertil Soils,* 49:791–801.

IFA (International Fertilizer Association). 2016. *Nitrogen Fertilizer Consumption Data.* Int. Fertilizer Industry Assoc., Paris, France. 12 Mar International Fertilizer Industry Association (IFA), Paris, France.

IPNI (International Plant Nutrition Institute). 2019. *4R Plant Nutrition Manual: A Manual for Improving the Management of Plant Nutrition.* http://www.ipni.net/ipniweb/portal/4r.nsf/ (accessed November 01, 2022).

Jakab, G., B. Madarász, M. Masoudi, M. Karlik, C. Király, D. Zacháry, T. Filep, I. Dekemati, C. Centeri, T. Al-Graiti, and Z. Szalai. 2023. Soil organic matter gain by reduced tillage intensity: Storage, pools, and chemical composition. *Soil and Tillage Research*, *226*:105584.

Jayne, T. S., N. J. Sitko, N. M. Mason, and D. Skole. 2018. Input subsidy programs and climate smart agriculture: Current realities and future potential. *Natural Resource Management and Policy*, 52:251–273.

Jin, J., Y. Fang, S. He, Y. Liu, C. Liu, F. Li, S. Khan, K. M. Eltohamy, B. Liu, and X. Liang. 2023. Improved phosphorus availability and reduced degree of phosphorus saturation by biochar-blended organic fertilizer addition to agricultural field soils. *Chemosphere*, *317:*137809.

Kelessidis, A., and A. S. Stasinakis. 2012. Comparative study of the methods used for treatment and final disposal of sewage sludge in European countries. *Waste Management*, *32*(6):1186–1195. https://www.epa.gov/nutrientpollution/sources-and-solutions-agriculture

Khan, M. N., M. Mobin, Z. K. Abbas, and S. A. Alamri. 2018. Fertilizers and Their Contaminants in Soils, Surface and Groundwater. In: Dominick A. DellaSala, and Michael I. Goldstein (eds.) *The Encyclopedia of the Anthropocene.* Oxford, England: Elsevier, 5:225-240.

Khan, A., Jan, M. T., Afzal, M., Muhammad, I., Jan, A., and Shah, Z. 2015. An Integrated Approach using Organic Amendments under a Range of Tillage Practices to Improve Wheat Productivity in a Cereal Based Cropping System. *International Journal of Agriculture and Biology*, *17*(3):467–474.

Kong, F., and S. Lu. 2022. Effects of microbial organic fertilizer (MOF) application on cadmium uptake of rice in acidic paddy soil: Regulation of the iron oxides driven by the soil microorganisms. *Environmental Pollution*, *307*, 119447.

Ladha, J. K., M. L. Jat, C. M. Stirling, D. Chakraborty, P. Pradhan, T. J. Krupnik, T. B. Sapkota, H. Pathak, D. S. Rana, K. Tesfaye, and B. Gerard. 2020. Achieving the sustainable development goals in agriculture: The crucial role of nitrogen in cereal-based systems. *Advances in Agronomy*, 163:39–116.

Li, R., R. Tao, N. Ling, and G. Chu. 2017. Chemical, organic and bio-fertilizer management practices effect on soil physicochemical property and antagonistic bacteria abundance of a cotton field: Implications for soil biological quality. *Soil & Tillage Research, 167*:30–38.

Lisboa, B. B., L. K. Vargas, A. O. Silveira, A. F. Martins, and P. A. Selbach. 2012. Indicadores microbianos de qualidade do solo em diferentes sistemas de manejo. Microbial indicators of soil quality in different management system.] *Revista Brasileira de Ciência Do Solo [Brazilian Journal of Soil Science], 36*(1):33–44.

Liu, L., A. Gunina, F. Zhang, Z. Cui, and J. Tian. 2022. Fungal necromass increases soil aggregation and organic matter chemical stability under improved cropland management and natural restoration. *Science of The Total Environment*, 858(3):159953.

Machado, D., T. R. Schossler, A. M. Zuffo, F. Andrade, and A. Piauilino. 2012. Atividades Microbianas E As Transformações No Ciclo Dos Elementos No Solo. [Microbial Activities and Transformations in the Cycle of Elements in Soil.] *Enciclopedia Biosfera [Biosphere Encyclopedia], 8*(15).

Maia, C. M. B. F., M. Guiotoku., R. T. G. Peixoto and L. M. P. Vargas. 2021. Biochar e eucalipto. [Biochar and eucalyptus.] In *O eucalipto e a Embrapa: quatro décadas de pesquisa e desenvolvimento [Eucalyptus and Embrapa: four decades of research and development],* ed. Oliveira, E. B and J. E. Pinto-Junior, 1160:589-609. Brasilia: Embrapa.

Maranguit, D., T. Guillaume, and Y. Kuzyakov. (2017). Land-use change affects phosphorus fractions in highly weathered tropical soils. *Catena*, 149, 385–393.

Marcińczyk, M and P. Oleszczuk. 2022. Biochar and engineered biochar as slow- and controlled-release fertilizers. *Journal of Cleaner Production*. 339:1-14.

Mateo-Marín, N., R. Isla, M. Guillén, and D. Quílez. 2020. Agronomic and Environmental Implications of Substituting Pig Slurry for Synthetic Nitrogen in Mediterranean Wheat Systems. *Agronomy, 10*(10):1498.

Mullen, M. D. 2019. Phosphorus in Soils—Biological Interactions. Reference Module in E*arth Systems and Environmental Sciences*.

Nascimento, C. A. C., G. C. Vitti, L. A. Faria, P. H. C. Luz, and F. L. Mendes. 2013. Ammonia volatilization from coated urea forms. *Revista Brasileira de Ciência Do Solo [Brazilian Journal of Soil Science], 37*(4):1057–1063.

Neal, A. L., A. Bacq-Labreuil, X. Zhang, I. M. Clark, K. Coleman, S. J. Mooney, K. Ritz, and J. W. Crawford. 2020. Soil as an extended composite phenotype of the microbial metagenome. *Scientific Reports, 10*(1): 1–16.

Noellsch, A. J., P. Motavalli., K. A. Nelson and N. R. Kitchen. 2009. Corn response to conventional and slow release nitrogen fertilizers across a clay plan landscape. *Agronomy Journal*, 101:607-614.

Nunes, R. de S., D. M. G. Sousa, W. J. Goedert, L. E. Z. Oliveira, P. S. Pavinato, and T. D. Pinheiro. 2020. Distribution of Soil Phosphorus Fractions as a Function of Long-Term Soil Tillage and Phosphate Fertilization Management. *Frontiers in Earth Science*, 8:350.

Olmo, R., S. U. Wetzels, J. S. L. Armanhi, P. Arruda, G. Berg, T. Cernava, P. D. Cotter, S. C. Araujo, R. S. C. Souza, I. Ferrocino, J. C. Frisvad, M. Georgalaki, H. H. Hansen, M. Kazou, G. S. Kiran, T. Kostic, S. Krauss-Etschmann, A. Kriaa, L. Lange, E. Maguin, B. Mitter, M. O. Nielsen, M. Olivares, N. M. Quijada, M. Romani-Perez, Y.

Sanz, M. Schloter, P. Schmitt-Kopplin, S. C. Seaton, J. Selvin, A. Sessitsch, M. Wang, B. Zwirzitz, E. Selbeeherr, and M. Wagner. 2022. Microbiome Research as an Effective Driver of Success Stories in Agrifood Systems – A Selection of Case Studies. *Frontiers in Microbiology*, *13*:2190.

Omara, P., L. Aula, F. B. Oyebiyi, E. M. Eickhoff, J. Carpenter, & W. R. Raun. 2020. Biochar Application in Combination with Inorganic Nitrogen Improves Maize Grain Yield, Nitrogen Uptake, and Use Efficiency in Temperate Soils. *Agronomy*, 10(9), 1241.

ONU, United Nations Organization. 2022. *Envisaging the Future of Cities*. World Cities Report 2022.

Östman, M., J. Fick, E. Nässtrõm, and R. H. Lindberg. 2014. A snapshot of illicit drug use in Sweden acquired through sewage water analysis. *The Science of the Total Environment*, *472*:862–871.

Pires, A. M. M. and Mattiazzo, M. E. 2008. *Avaliação da viabilidade do uso de resíduos na agricultura.* [*Evaluation of the feasibility of the use of residues in agriculture.*] Jaguariúna, São Paulo: *Embrapa Meio Ambiente*, 1-9.

Prates, A. R., A. R. Coscione, M. C. M. Teixeira Filho, B. G. Miranda, O. Arf, C. H. Abreu Junior, F. C. Oliveira, A. Moreira, F. S. Galindo, M. M. P. Sartori, Z. He, A. D. Jani, G. F. Capra, A. Ganga, and T. A. R. Nogueira. 2020. Composted sewage sludge enhances soybean production and agronomic performances in naturally infertile soils (Cerrado region, Brazil). *Agronomy-Basel* 10(11):1-20.

Qian, L., B. Chen, and D. Hu. 2013. Effective alleviation of aluminum phytotoxicity by manure-derived biochar. *Environmental Science & Technology*, *47*(6):2737–2745.

Redding, M. R., R. Lewis, T. Kearton, and O. Smith. 2016. Manure and sorbent fertilisers increase on-going nutrient availability relative to conventional fertilisers. *The Science of the Total Environment*, *569–570*:927–936.

Robertson, G. P., and P. M. Vitousek. 2009. Nitrogen in Agriculture: Balancing the Cost of an Essential Resource. *Annual Review of Environment and Resources*, 34:97–125.

Rós, A. B., N. Narita, A. C. H. Silva, and J. E. Creste. 2020. Effects of limestone and organic fertilizer on cassava yield and on chemical and physical soil properties. *Revista Ceres*, *67*(1):23–29.

Said, A., Q. Zhang, J. Qu, Y. Liu, Z. Lei, H. Hu, and Z. Xu. 2018. Mechanochemical activation of phlogopite to directly produce slow-release potassium fertilizer. *Applied Clay Science*, 165:77–81.

Saleem, K., M. A. Asghar, M. H. Saleem, A. Raza, G. Kocsy, N. Iqbal, B. Ali, M. F. Albeshr, and E. A. Bhat. 2022. Chrysotile-Asbestos-Induced Damage in Panicum virgatum and Phleum pretense Species and Its Alleviation by Organic-Soil Amendment. *Sustainability, 14*(17):10824.

Smith, W. B., M. Wilson, P. Pagliari. 2020. Organomineral fertilizers and their application to field crops. *Animal Manure: Production, Characteristics, Environmental Concerns, and Management*, 67:229-243.

Solgi, E., H. Sheikhzadeh, and M. Solgi. 2018. Role of irrigation water, inorganic and organic fertilizers in soil and crop contamination by potentially hazardous elements in intensive farming systems: Case study from Moghan agro-industry, Iran. *Journal of Geochemical Exploration*, 185:74-80.

Souza, T. L., D. R. Guelfi., A. L. Silva., A. B. Andrade., W. F. T. Chagas and E. L. Cancellier. 2017. Ammonia and carbon dioxide emissions by stabilized conventional nitrogen fertilizers and controlled release in corn crop. *Revista Ciência e Agrotecnologia [Science and Agrotechnology Journal]*, 41:494-510.

Spark Strategic Intelligence (SSI). 2021. O future é biologico. [The future is biological.] *Spark smart decisions*. Available from: < http://spark-ie.com.br/>. Accessed: Nov 20. 2022.

Suter, H. C., P. Pengthamkeerati, C. Walker, & D. Chen. 2011. Influence of temperature and soil type on inhibition of urea hydrolysis by N-(n-butyl) thiophosphoric triamide in wheat and pasture soils in south-eastern Australia. *Soil Research*, 49(4), 315.

Taurus. 2022. *Preventing nitrogen loss in your fall applications*. Available at:< https://taurus.ag/preventing-nitrogen-loss-in-your-fall-applications/> Accessed on: November 20, 2022.

Tejada, M., I. Gómez, E. Fernández-Boy, and M. J. Díaz. 2014. Effects of Sewage Sludge and Acacia dealbata Composts on Soil Biochemical and Chemical Properties. *Communications in Soil Science and Plant Analysis*, *45*(5):570–580.

Thomas, C. L., G. E. Acquah, A. P. Whitmore, S. P. McGrath, and S. M. Haefele. 2019. The Effect of Different Organic Fertilizers on Yield and Soil and Crop Nutrient Concentrations. *Agronomy, 9*(12):776.

Trenkel, M. 2010. Definitions of slow- and controlled-release fertilizers and nitrification and urease inhibitors. In *Slow- and controlled-release and stabilized fertilizers: an option for enhancing nutrient efficiency in agriculture*, ed Trenkel, M, 1:14-20. Paris: International Fertilizer Industry Association.

USEPA (United States Environmental Protection Agency). 1999. *Environmental Regulations and Technology*. https://www.epa.gov/sites/default/files/2015-04/documents/control_of_pathogens_and_vector_attracton_in_sewage_sludge_july_2003.pdf. (accessed November 20, 2022).

USEPA (United States Environmental Protection Agency). 2003. T*est Methods for Evaluating Solid Residue*, Physical/Chemical Methods, EPA publication SW 846, Third Edition, Final Updates I (1993), II (1995), IIA (1994), IIB (1995), III (1997), IIIA (1999), IIIB (2005), IV (2008), and V (2015).

USEPA (United States Environmental Protection Agency). 2022. T*he Sources and Solutions: Agriculture*. https://www.epa.gov/nutrientpollution/sources-and-solutions-agriculture (accessed October 26, 2022).

USEPA (United States Environmental Protection Agency). (2022a). *Basic Information about Biosolids*. https://www.epa.gov/biosolids/basic-information-about-biosolids (accessed October 25, 2022).

Valle, S. F., A. S. Giroto., G. G. F. Guimarães., K. A. Nagel., A. Galinski., J. Cohnen., N. D. Jablonowski and C. Ribeiro. 2022. Co-fertilization of Sulfur and Struvite-Phosphorus in a Slow-Release Fertilizer Improves Soybean Cultivation. *Front. Plant Science*, 13:1-12.

Van Raij, B. 2011. *Fertilidade do solo e manejo de nutrientes*. [*Soil fertility and nutrient management*.] Instituto Internacional de Nutrição Vegetal, Piracicaba, Brasil.

Verburg, Kirsten. 2020. Understanding soil water effects on nitrogen release from controlled-release fertilizers. *Soil Science Society American*, 85:59-72.

Verburg, Kirsten. 2022. *What are controlled-release fertilizers?* Available at:< https://soilsmatter.wordpress.com/2022/03/15/what-are-controlled-release-fertilizers/> Accessed: November 22, 2022.

Wang, Z., M. Xue, Y. Wang, M. Song, S. Li, R. A. Daziano, B. Wang, G. Ma, K. Chen, X. Li, B. Zhang. 2019. Big data: new tend to sustainable consumption research. *Journal of Cleaner Production,* 236:117499.

Wei, X., J. Chen, B. Gao, & Z. Wang. 2020. Role of controlled and slow release fertilizers in fruit crop nutrition. *Fruit Crops*, 555–566.

Weifeng, S., A. Shu, J. Liu, W. Shi, M. Li, W. Zhang, Z. Li, G. Liu, F. Yuan, S. Zhang, Z. Liu, and Z. Gao. 2022. Effects of long-term fertilization with different substitution ratios of organic fertilizer on paddy soil. *Pedosphere*, 32(4):637-648.

Wu, L., Y. Jiang, X. He, H. Liu, and K. Yu. 2020. Increased organic fertilizer application and reduced chemical fertilizer application affect the soil properties and bacterial communities of grape rhizosphere soil. *Scientific Reports*, 10(1):1-10.

Ye, L., M. Camps-Arbestain., Q. Shen., J. Lehmann., B. Singh and M. Sabir. 2020. Biochar effects on crop yields with and without fertilizer: A meta-analysis of field studies using separate controls. *Soil Use Manage*. 36:2– 18.

Zhang, Y., W. Gao, L. Ma, H. Luan, J. Tang, R.. Li, M. Li, S. Huang, and W. Lei. 2023. Long-term partial substitution of chemical fertilizer by organic amendments influences soil microbial functional diversity of phosphorus cycling and improves phosphorus availability in greenhouse vegetable production. *Agriculture, Ecosystems & Environment*, 341:108193.

Zonta, E., J. B. Stafanato, M. G. Pereira. 2021. Fertilizantes minerais, orgânicos e organominerais. *Embrapa Mandioca e Fruticultura-Capítulo em livro técnico* (INFOTECA-E).

Chapter 4

Soil Aggregation: Formation, Stabilization, and Impacts of Soil Management

Maurício Cunha Almeida Leite[1,*]
Thaís Nascimento Pessoa[1]
Maria da Costa Cardoso[2]
David Lukas de Arruda Silva[2]
and Daniel Nunes da Silva Júnior[2]
[1]Department of Soil Science, University of São Paulo,
"Luiz de Queiroz" College of Agriculture, Piracicaba, São Paulo, Brazil
[2]Department of Soil Science, Federal University of Viçosa, Viçosa, Minas Gerais, Brazil

Abstract

Soil structure has as its main component the soil aggregate, which is composed primarily of particles of sand, silt, and clay, bound together by inorganic and organic materials at various scales. However, the concepts and models explaining how soil aggregates form and stabilize have evolved over the years. Nowadays, it is understood that microaggregates are primally formed from the disruption of macroaggregates. Nevertheless, free microaggregates can feed back into the system to form new macroaggregates. During this process, physical-chemical interactions such as flocculation, organic bonds, and inorganic cementing agents such as iron and aluminum oxyhydroxides give rise to soil microaggregates. The main methods utilized to evaluate the stability of the aggregates can be grouped by (i) the amount of clay dispersed (WDC); (ii) methods that evaluate soil aggregates by size classes or

* Corresponding Author's Email: cunhamauricio@usp.br.

In: Agricultural Soils
Editors: R. Nogueira de Sousa, T. Nascimento Pessoa and K. Ferreira Mendes
ISBN: 979-8-89113-000-5

indices such as mean weight diameter (MWD) and geometric mean diameter (GMD) or (iii) by evaluating the tensile strength (TS). Organic factors are also undoubtedly extremely relevant to the formation and stabilization of soil aggregates. Soil ecosystem engineers as earthworms and termites mediate the formation of soil structure and aggregation. Earthworms can act as catalysts for the decomposition of soil organic matter by moving the soil matrix during the burrowing process. Moreover, earthworms produce mucus, and their guts and casting process mix soil, organic particles, and binding agents in the same process. On the other hand, termites are known to increase the formation of soil organic matter and the activity of microorganisms, as well as produce casts and mucus as binders for soil aggregation formation and stabilization. Furthermore, soil organic matter and the microorganisms, which are ubiquitous in most soils act mainly in the formation of macroaggregates and, consequently, influence the structure and porosity of the soil. The roots' action is due to the exudation of extracellular polymeric substances (EPS) and to the physical association of individual particles. Microorganisms act in a similar manner. Bacteria can secrete substances onto the colloidal surface that acts as a binding agent, or fungal hyphae can physically bind particles together. Soil aggregation is one of the parameters used to assess Soil Physical Quality (SPQ), an important component used to assess Soil Quality (SQ). In this way, it is possible to evaluate different land uses and soil management systems in terms of their quality. Soil aggregation is relatively sensitive to soil management because of the wide range of processes and mechanisms that are involved in the formation and stabilization of soil aggregation.

Keywords: soil structure, macroaggregates, microaggregates, binding agents, land use, soil quality

Introduction

Soil aggregates are developed from individual particles (i.e., sand, silt, and clay). Moreover, organic compounds and inorganic cements are required to bind these particles together to form soil aggregates. However, the soil aggregation process occurs at different scales and degrees of cohesion and is sensitive to the content and quality of the organic material involved, the meso and microfauna, and the soil mineralogy. The sum of these components in various processes and mechanisms determines the stability of the aggregates (Blanco-Canqui and Lal, 2004). Then, the organization of these aggregates forms the soil pore space. Thus, the solid phase and soil pore space (containing water and air) develop the soil structure (Tisdall, 2020). Therefore, soil structure combines processes and mechanisms among the physical, chemical, and biological properties of the soil.

Soil aggregation is based on different processes and mechanisms at different scales and the studies and theories behind this process have changed over time. After identifying the physical and chemical processes that govern soil aggregation in the early 20th century, soil scientists began to develop models and frameworks to explain how soil aggregation occurs (Six et al., 2004). As well as the evolutionary question "who came first, the chicken or the egg?"; the models tried to explain how soil micro and macroaggregates should be formed and their influence on each other. Emerson (1959) pioneered the conceptualization of soil aggregate formation in a model linking soil organic matter (SOM) and soil aggregation. After 1980, the concept of hierarchy was elaborated and added to models of aggregate formation and stabilization (Tisdall and Oades, 1982). However, Oades (1984) postulated an opposite chronology in the hierarchy of soil aggregation formation, in which the origin of the microaggregates would come from the breakdown of the macroaggregates. Nevertheless, the hierarchy concept can be applied to soils where SOM is the major binding agent (Oades and Waters, 1991). Subsequently, researchers are attempting to elucidate the processes and mechanisms involved in soil aggregation, with an emphasis on the environment. For example, topics such as the role of soil fauna (Shipitalo and Protz, 1989), microorganisms' activity on soil aggregates (Bucka et al., 2021), soil carbon stock on soil and aggregate turnover (Plante and McGill, 2002a,b) or soil aggregation as a qualitative assessment for soil quality assessment (Bünemann et al., 2018) have been highlighted.

The main factors influencing soil aggregation can be divided into two groups: inorganic and organic. The inorganic group consists of interparticle forces such as van der Waals and electrostatic, flocculation and dispersion phenomena. Mineralogical differences between 1:1, 2:1 layers or iron and aluminum oxide clays can also play an important role in aggregate formation. On the other hand, organic factors (OM), consisting of organic matter fractions and soil fauna, are the main pillar of the soil aggregation process. Soil macro-, micro-, and mesofauna are responsible for the decomposition and movement of soil organic matter between soil layers. Earthworms are an example of macrofauna that can act as a catalyst in this system, moving OM from the soil surface to deep layers. In addition, soil microfauna such as soil fungi and bacteria act mainly as soil binders, breaking down OM into small fractions as particulate organic matter (POM) and/or exuding polysaccharides and complex substances during their life cycle. As a result, fine organic matter and fungal hyphae act as binding agents for the formation of microaggregates. The degradation of this free organic matter leads to the formation of macroaggregates by binding microaggregates into macroaggregates.

Soil aggregation can be quantified and assessed by a variety of techniques, such as clay dispersed in soil (WDC), distribution of soil aggregates in different size classes by diameter (MWD), and even by their tensile strength (TS). The smaller the scale of the processes, the more sensitive the process will be to environmental impacts. In this way, different parameters can be used to determine or estimate soil quality. Studies show that soil use and management have a major impact on soil aggregation, and consequently on the cycling of organic matter and the diversity of soil fauna. Because soil aggregation involves all components of the soil environment, it should play a key role in future agricultural scenarios where soil quality and health will be an ecological and social necessity.

History of Soil Aggregation and Stabilisation Research

Studies and observations on soil aggregate formation and stabilization are not recent. In fact, all the main factors that govern this process were identified in the early 1900s. In this period, it was possible to identify that soil organic matter as roots, soil fauna, soil microorganisms, inorganic binding agents, and other variables such as soil type, water content, and climate would be the main factors responsible for soil aggregation. However, it was only after the 1950s that research began to try to develop models and frameworks about the genesis and stabilization of soil aggregates. The objective during this period was to describe how the process would occur, dividing it into stages and actions of each main agent.

The work of Martin et al., (1955) provides a summary of the influence and interaction of the main factors in aggregation. In this review, the influence of organic residues, microbial activity, cation exchange, and surface loading on soil aggregation was already discussed. Subsequently, the first scientific work that stands out for the conceptualization of such ideas was that of Emerson (1959), who proposed a conceptual model for the primary relationship between soil organic matter (SOM) and soil aggregates. In his model, soil aggregation could be formed by oriented clay layers and quartz particles bounded by SOM. He was the first to deduce a stratified model of forming. Subsequently, Edwards and Bremner (1967) proposed the next theoretical model to be highlighted. They presented a theory focused on the formation of microaggregates, conceptualizing their formation through the reaction of a solid phase as clays (C) and polyvalent metals (P) between soil organic matter, in a structure called C-P-SOM each. Although Emerson's (1959) models appear similar, Edwards and Bremner (1967) also proposed C-P-C and SOM-P-SOM, suggesting different pathways. In addition, they started the discussion on the complexation of organic matter during the formation of microaggregates and how this could provide physical protection against degradation factors such as microorganisms. They also concluded that soils are a matrix of microaggregates (<250μm) and their association to form macroaggregates (>250μm).

By adding more detail to the soil aggregation model, Tisdall and Oades (1982) placed great emphasis on the concept of soil aggregation. They showed the interactions between the aggregates and the SOM over the time of action of each of the binding agents in the process of aggregate formation. At this time, the concept of hierarchy and transient, temporary, and persistent binders was introduced to build a more complex model. Their work postulated that the binding between free primary particles and silt-size aggregates (<20-50 μm) involves organic and inorganic persistent binders such as humified organic matter, polyvalent cation complexes, oxyhydroxides, and aluminosilicates. Particles even smaller than 20 μm are also organized into structural units containing minerals, SOM, as well as microbial biomass, DNA, RNA, proteins, and carbohydrates (Plante and McGill, 2002; Tisdall and Oades, 1982), and can therefore be defined as aggregated. According to Totsche et al. (2018), microaggregates can be defined as structures that are smaller than 250 μm and are composed of various mineral, organic, and biotic materials that act as a nucleus or a cementing agent. As a result, stable microaggregates can coalesce into macroaggregates. Macroaggregates, in turn, include all soil aggregates (>250μm) through mainly temporary binding agents such as fungal hyphae and/or roots. In addition, smaller microaggregates (<50mm) should also include transient binders (i.e., microbial and plant exudates in the form of polysaccharides).

Continuing his research on soil aggregates, Oades (1984) formulated a major modification of the previously proposed model. His earlier work (Tisdall and Oades, 1982) proposed a linear hierarchical model from the smallest to the largest aggregate. However, Oades (1984) postulated an opposite chronology in the hierarchy of soil aggregate formation. In this case, Oades (1984) hypothesized that the origin of microaggregates would come from the breakdown of macroaggregates. This process should happen because roots and hyphae, binding agents of macroaggregates, would form the microaggregates nucleus, inside macroaggregates. The justification for this process is since roots and hyphae are temporary binding agents, during their decomposition, mucilages are produced, forming encrustations on the mineral faces inside the macroaggregates. This process then gives rise to microaggregates within macroaggregates and after some environmental stress, these macroaggregates are broken down into smaller fractions, releasing microaggregates into the soil environment (Figure 1).

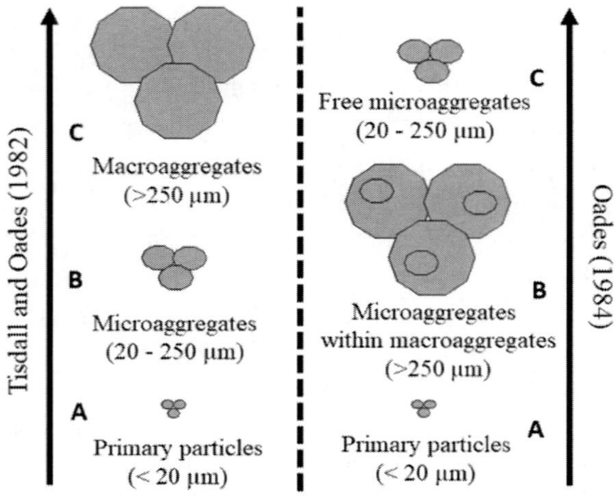

Source: Adapted from Six et al., (2004).

Figure 1. Hierarchical conceptual model of soil aggregate formation and stabilization by Tisdall and Oades (1982) in comparison with the model postulated by Oades (1984).

The model of Oades (1984) was later explored by Angers et al., (1997), who presented quantitative data to support the model of microaggregates within macroaggregates. Their work observed changes in the redistribution of carbon (C) and nitrogen (N) isotopes through soil aggregates in agricultural systems. The work was carried out by labeling ^{13}C and ^{15}N in macro- and microaggregates during the decomposition of wheat straw. In this case, the data showed an increase of ^{13}C in macroaggregates during the initial decomposition of the straw. However, as the decomposition stabilized, there was a decrease in ^{13}C levels in macroaggregates and a stabilization of ^{13}C in microaggregates. These quantitative observations of C redistribution then helped to demonstrate more clearly that there was indeed the formation of microaggregates within macroaggregates and their release after macroaggregate breakdown.

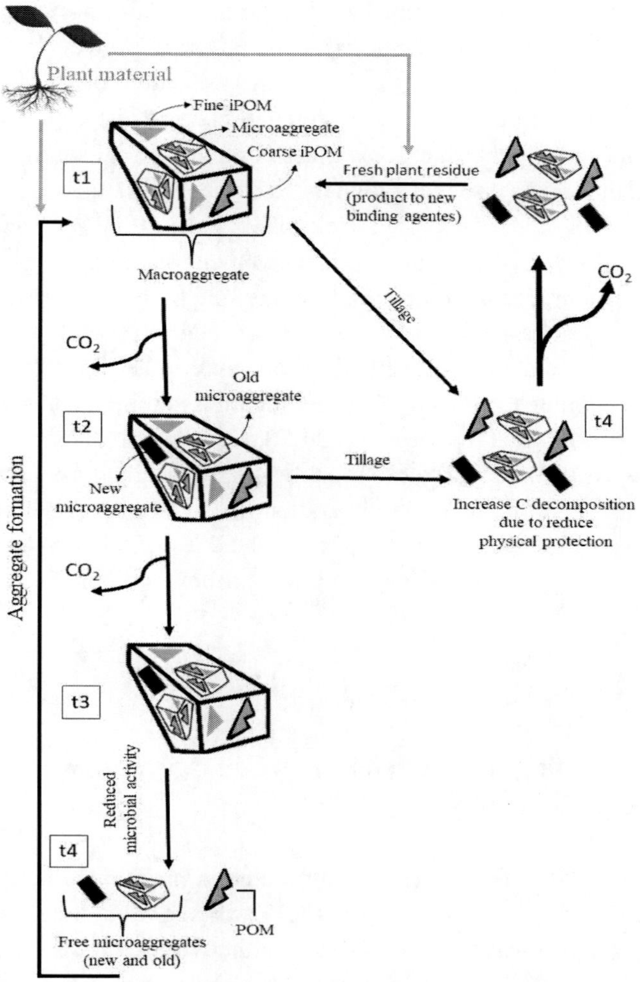

Source: Six et al., (2000).

Figure 2. Conceptual model of macroaggregate "life cycle" and microaggregate formation. Factors act in the process between fresh organic matter input, SOM, POM, macroaggregate turnover and SOM and iPOM (intra-aggregate particulate organic matter). In the absence of disturbance, turnover occurs during aggregate formation, stabilization and destabilization. On the other hand, factors that cause abrupt changes in the system, such as tillage, reduce the life cycle of macroaggregates.

Since 2000, developments have been made to understand and improve models of soil aggregation formation and stabilization. Plante and McGill, (2002a,b) observed the relationship between the occlusion of organic matter and the turnover of macroaggregates as physical protection in a tillage situation. They propose that the physical protection of MOS and POM resulting from the formation of soil aggregates can be altered by newly added C inputs or, on the other hand, by excess output from C sources. They also suggested that soil C sequestration is maximal when there is a moderate turnover of soil aggregates. The work of Six et al., (2004) provides a more detailed overview of the history and development of soil aggregation studies. In addition, more recent research has already sought to use soil aggregation as a qualitative assessment, called structural stability, within soil quality assessment (Bünemann et al., 2018). Meanwhile, other lines of research are already attempting to quantify soil aggregation stability using visible-near infrared spectroscopy (Afriyie et al., 2022).

Currently, the research seeks to increasingly integrate assessments of the main factors of formation and stabilization in soil aggregation studies. While studies are also carried out on the effect of land use and land management on this land property, as will be observed in the following topics.

Factors Affecting Soil Aggregation

Inorganic Binding Agents Involved in Soil Aggregation and Structural Stability

The sedimentation of particles is controlled and maintained by gravitational force. This force is generated by the Earth's gravitational field and is negligible when considering the interaction between colloids (Bleam, 2016). Solid particles suspended in a liquid settle under gravitational force. This settling is estimated by the sedimentation rate, which is calculated using Stokes' law and depends on the buoyancy of the particles, the viscosity of the liquid, and the size of the particles (Dahlgren et al., 2008). In addition, intrinsic forces such as attraction and repulsion act between colloids, attracting and repulsing them respectively (Sposito, 2008).

Suspended clays are subject to interparticle forces (Chibowski, 2011). Attractive intermolecular forces, which include van der Waals and electrostatic forces, occur when the particles are very close to each other and are a result of the dipole orientation or the coordination of the electron motion between the particles (Sposito, 2008). The van der Waals force is only effective at very close distances and decays rapidly with increasing distance. The repulsive forces, which include the electrostatic force, are a consequence of the repulsion of electric charges of the same sign and are more active at greater distances (Tan, 2010).

In soil, clays are commonly negatively charged and balanced by exchangeable cations adsorbed on their surfaces. (Melo and Alleoni, 2009). In suspension, cations tend to bind on the surface of clay minerals to balance the concentration difference between the interface and the liquid phase. (Sposito, 2008). However, most of these ions, especially those near the clay surface cannot move very far due to the strong attraction of the negative charge on the clay surface (Tan, 2010). Cations aggregate at the interface, forming an electrical double layer, which can vary in thickness from 50 to 300 Å (Tan, 2010). Whenever these clay particles come together, repulsion occurs between the particles due to the outer segment of the double layers having the same type of (positive) charges (Sposito, 2008). The suspension is considerably stable, and the clay remains dispersed.

The phenomena of dispersion and flocculation influence the stability of the aggregates. Aggregation mechanisms can be summarized in Figure 3 where *A* represents aggregation processes and *D* dispersion (Lal and Shukla, 2004). The water-dispersible clay depends on the stability of the aggregates in water, the number of exchangeable cations (e.g., Al^{3+}, Ca^{2+}, Mg^{2+}, K^{+}), the presence of complexing agents, silicate minerals, iron/aluminum oxides, ionic strength and pH (Alleoni et al., 2009). Carbonates, sesquioxides, aluminosilicates, and hydrated silicates are the most important inorganic cementing agents (Harris et al., 1966).

Soil dispersion and flocculation processes depend on the behavior of the diffuse double layer (DDL). The DDL is also called the electric double layer. The flocculation of clays occurs preferentially in the presence of polyvalent ions with a small hydration radius. These ions promote the compression of the DDL. The ions adsorbed on the surface of the colloid can compress the DDL and promote the flocculation process because the electrical potential of the surface is close to zero. Usually, less weathered soils can have low structural stability due to the presence of mono- and divalent cations in the exchange complex (Ca^{2+}, Mg^{2+}, Na^{+}, K^{+}). These cations cause the extension of the DDL beyond the Stern layer, promoting a high electric potential, factors that increase the dispersion of the colloids. On the other hand, in highly weathered soils, the exchange complex is composed of Al^{3+}, which compresses the DDL and increases flocculation (Alleoni et al., 2009).

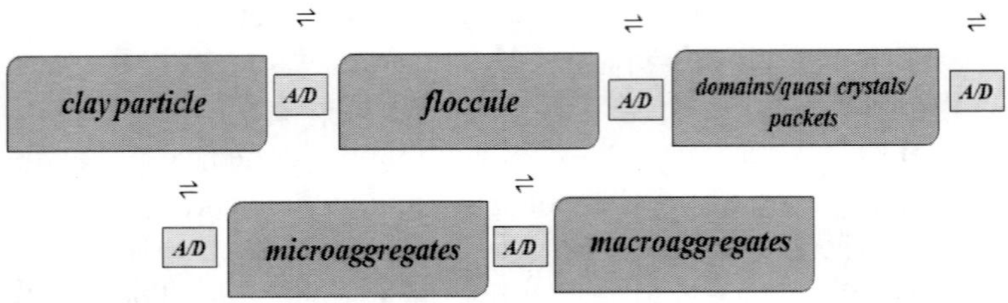

Source: Adapted from Lal and Shukla (2004).

Figure 3. Aggregation mechanisms, where *A* represents aggregation processes and *D* dispersion.

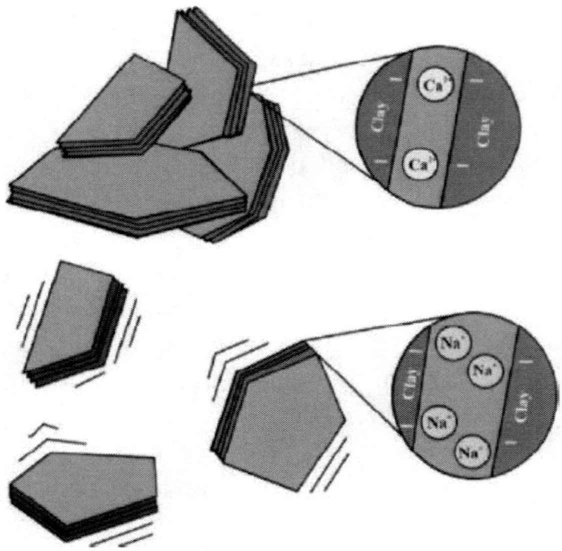

Figure 4. The di- and trivalent cations such as Ca^{2+}, Fe^{3+}, and Al^{3+} are strongly sorbed, form bridges that unite clay particles, and effectively neutralize negative charges on clay surfaces. A negative charge balance can cause clay particles to repel and create a dispersing condition. (1) the large hydrated Na^+ ions cannot get close enough to effectively neutralize the negative charges, (2) the single charge on Na^+ is not effective in bridging the clay particles.

It is well known that clay flocculates when the pH is close to the point of zero charge (PZC); if the pH remains below the PZC, there will be a predominance of positive charges, and if it is above the PZC, negative charges will predominate. Thus, this predominance of net negative or positive charges, depending on the pH of the soil, leads to repulsive forces that are strong enough to overcome the electrostatic forces of attraction, in which case much of the clay remains dispersed (Gillman, 1974).

Substitution of Na^+ by cations such as K^+, Ca^{2+}, and Mg^{2+} at the edges of clay minerals also controls flocculation by changing the thickness of the DDL (Figure 4) (Fontes et al., 2001). The dispersion of colloids is therefore related to the interaction of electrical charges on their surface in a polar solvent; these electrical charges can be generated by isomorphic substitution or by dissociation of radicals at the edges of mineral structures. The electrostatic charge generated by isomorphic substitution is not variable, whereas that generated by the dissociation of OH radicals vary with the chemical characteristics of the soil, such as the pH and electrolyte concentration of the soil solution (Theng, 1980; Yu, 1997).

Soil mineralogy, particularly clay-sized minerals, influences properties that affect aggregation: surface area, cation exchange capacity (CEC), charge density, dispersity, and expansibility (Bronick and Lal, 2005). Clay minerals occur as layered crystalline structures or as non-crystalline minerals with an amorphous structure (Dixon and Schulze, 2018). Non-crystalline clay minerals, such as allophanes and imogolites, have high surface areas and pH-dependent charge properties that generally increase aggregation (Powers and Schlesinger, 2002).

Non-expansive clay minerals, such as kaolinite (1:1), have low CEC and surface area, which tends to reduce aggregate stability. Kaolinite has a good flocculation capacity due to electrostatic charges between mineral plates (Chibowski, 2011). Highly active clays such as smectites and other 2:1 clays, which are associated with high CEC and large specific surface areas, generally have high aggregation when compared to 1:1 minerals (Seta and Karathanasis, 1996). The expansive nature of smectites (Bartoli et al., 1988; Piccolo et al., 1997) can cause aggregates to break up during wetting and drying cycles.

Iron (Fe) and aluminum (Al) oxides are also important inorganic cementing agents that stabilize soil aggregates. These oxides are particularly important in highly weathered soils where the clay fraction is predominantly composed of Fe and Al oxides and kaolinite (Schaefer et al., 2008; Pessoa et al., 2022). It is important to note that there are different mechanisms and binders for macroaggregates and microaggregates, controlled for organic and inorganic compounds. Macroaggregates are stabilized and modified by changes in the quantity and quality of soil organic matter (SOM) and by soil use and management (Amelung et al., 2023). The formation of microaggregates is part of the process of pedogenesis and the main binders are organo-mineral complexes. The main cementing agents in microaggregates are Fe, Al, Mn and other soil organic binders. The main minerals in microaggregate formation are Fe and Al oxides (Totsche et al., 2018). Amelung et al., (2023) observed in an experiment that the organic binders initiated macroaggregate formation and had greater relevance compared to the mineral. The Fe oxides and the argilominerals added to the soil participated in all the aggregate formation processes in a similar way, there was no preference for participation in a specific size. It was possible to observe that the labile organic materials suffered degradation, while the added minerals continued to participate in the cementation of the aggregates, as changes occurred in the occluded and stable fractions (Figure 5). Thus, the study showed an aggregate formation that follows a temporal sequence in which the organic substrates initiate the process and the minerals stabilize the process in the long term.

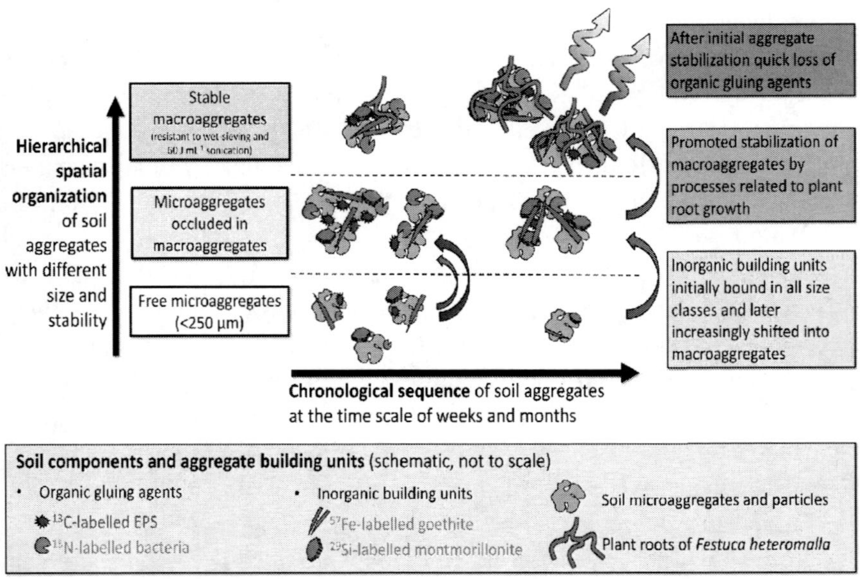

Source: Amelung et al., (2023).

Figure 5. The spatial organization of microaggregates and macroaggregates is a product of organic binding agents and inorganic building units. The formation of aggregates responds to a chronological sequence.

Organic Factors in the Soil Aggregation Process

Soil quality can be defined as "the potential of a soil to function within ecosystem and land-use boundaries to sustain biological productivity, maintain environmental quality and promote plant and animal health" (Doran and Parkin, 1996, 1994). Soil sustainability and resilience also derive from the soil's potential to self-regulate and resist changes in its environment. These fundamental soil properties are developed by the main factors of soil formation: parent material, climate, topography, organisms and time. Except for time, the soil-forming factors are considered to be interdependent, with multiple response effects occurring between them (Lavelle et al., 1993). The soil fauna has a small-scale activity potential relative to other processes in the soil ecosystem (Figure 6) and therefore its role in the cycling of elements such as C is fundamental.

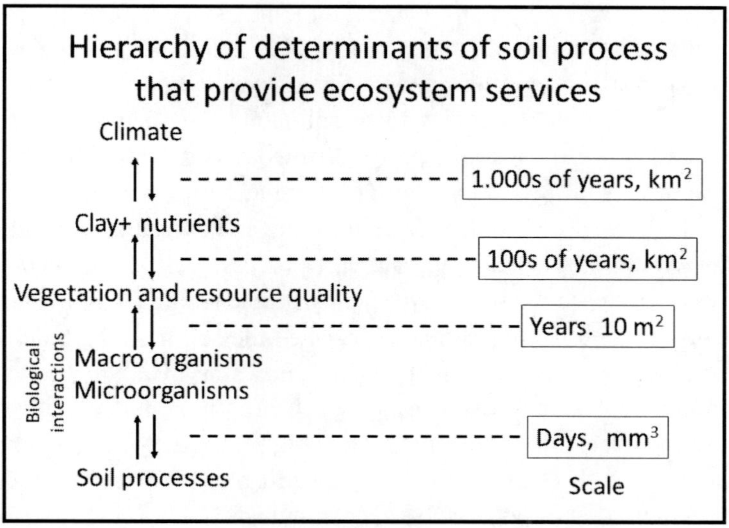

Source: Adapted from Lavelle et al., (1993).

Figure 6. Potential time scale contribution hierarchy of different factors acting on the soil ecosystem.

The need for increased food production and inappropriate land management are making our soils more vulnerable to soil degradation and the loss of soil biodiversity. These issues make it increasingly important to understand the influence of soil fauna and microorganisms on soil aggregation. Soil formation and stabilization have always been viewed as processes driven by biota pathways (Edwards and Bohlen, 1996; Martin, 1991; Tisdall and Oades, 1982). Soil macro- and microfauna play an essential role in soil aggregation and porosity through their burrowing and mixing activities. Some aspects and examples of soil fauna and micro-organisms are discussed below in relation to their role in soil quality and aggregation.

Soil Ecosystem Engineers: Earthworms and Termites

The first protagonist of studies on the effect of soil fauna on the soil itself was the earthworm. Darwin was the first researcher to mention the functions of earthworms in the ecosystem through his observations. The works of Edwards and Bohlen (1996), Lee and Foster (1991), and Martin (1991) describe the role of earthworms in transporting litter and other organic sediments from the soil surface and thus incorporating them into the development chain of soil aggregates. The origin of macroaggregates can be distinguished from physicogenic or biogenic, which are formed by the binding of soil minerals during wetting and drying cycles, or by soil fauna and microorganisms.

According to Brown et al. (2000), earthworms can play different roles in the process of soil aggregation by (1) moving and mixing the soil matrix through the burrowing process or (2) ingesting and excreting soil and organic particles through the mucus, gut and cast process. Burrowing occurs in two main ways: (a) when the soil is sufficiently friable to allow earthworms to dig it without the need for ingestion, or (b) when the soil has a high density and ingestion by peristaltic movement becomes necessary. Both pathways lead to the formation of so-called burrows, which are an important factor in the structure and formation of soil aggregates (Liu et al., 2019). On the other hand, the earthworm body and digestion have specific mechanisms that produce and secrete mucus, a set of polysaccharide compounds (Guhra et al., 2020; Scheu, 1991). The mucus can act as a binder and substrate for microbial biomass (Martin J. Shipitalo and Le Bayon, 2004).

Furthermore, it is interesting to note that only certain earthworm species influence aggregation formation and SOM dynamics. Bouché, (1983) studied the establishment of earthworm communities and divided them into three main categories: (1) epigeic species (e.g., Lumbricus rubellus); (2) anecic species (e.g., Lumbricus terrestris); and (3) endogeic species (e.g., Lolobophora chlorotica). Epigeic species, known as litter dwellers, are small and live between the O and A horizons of the soil. The behavior of this group above the soil surface has no direct effect on soil structure or aggregation (Six et al., 2004). Anecic species live in deeper vertical burrows and come to the surface to feed and mate. They may bring organic debris such as leaves from the soil surface into their burrows (Hallam and Hodson, 2020). Endogeic species live in horizontal burrows in the upper soil layers and feed mostly on soil enriched with organic matter. They are considered the most important category in terms of soil aggregation and SOM stabilisation due to their feeding behaviour and the soil layer in which they act (Shipitalo and Bayon, 2004).

In addition to their contribution to soil aggregation and stabilisation, earthworms also play an important role in soil carbon storage and stabilisation. Mccartney et al., (1997) describe that after 3 years of reducing the earthworm population by electroshock, there was an 18% increase in soil carbon. Lavelle and Martin, (1992) described the effect of endogeic tropical earthworm gut transit in accelerating the decomposition of SOM. Bossuyt et al., (2004) investigated the effect of earthworms or their absence on the formation of stable microaggregates within macroaggregates and the distribution of C in them. They observed a significant increase (four times greater) in newly protected C in microaggregates within macroaggregates (>2000 μm) in the presence of earthworms. Changes in C stocks could be related to the potential of earthworms to increase decomposition rates by intestinal or even mucus association (Bossuyt et al., 2004; Martin, 1991; Serniak, 2017). On the contrary, earthworms can increase C stocks by long-term stabilisation within microaggregates, especially their gut microbiota (Zhang et al., 2013), or even by stabilization of soil microbial necromass (Zhang et al., 2013).

Through several studies (Ojha and Devkota, 2014; Bossuyt et al., 2005, 2004; Guhra et al., 2021, 2020; Lee and Foster, 1991; Shipitalo and Bayon, 2004; Shipitalo and Protz, 1989), earthworms have been found to play a primordial role in increasing SOM processing and microorganism activity, as well as being able to produce cast and mucus as binding agents for soil aggregation formation and stabilization. Therefore, studies indicate that earthworms play an important role in the soil ecosystem as catalysts for soil aggregation and C stabilization in microaggregates.

Termites are social insects and therefore live in communities (Lee and Foster, 1991). They act as soil bioturbators, resulting in the movement of material in the soil (Jones et al., 1994). This action in turn drives many other processes, transporting organic matter and nutrients and regulating ecological functions (Müller-Lemans and van Dorp, 1996; Wright and Jones, 2006). There are many reviews of the previous literature on the effects of termites on soils (Abe et al., 2000; De Bruyn and Conacher, 1990; Harit et al., 2017; Holt and Lepage, 2000; Lee and Wood, 1971). No wonder there is a growing interest in the study of soil engineers that can carry out this process, as in this case termites. They use soil from various depths, select specific soil particles by size for nest building above or below ground, and move a variety of organic and soil materials in tunnels (Jouquet et al., 2010). For example, foraging termites relocate and concentrate plant material, litter, and size-particle minerals (containing adsorbed nutrients such as nitrogen and phosphorus) in their mounds (Figure 7).

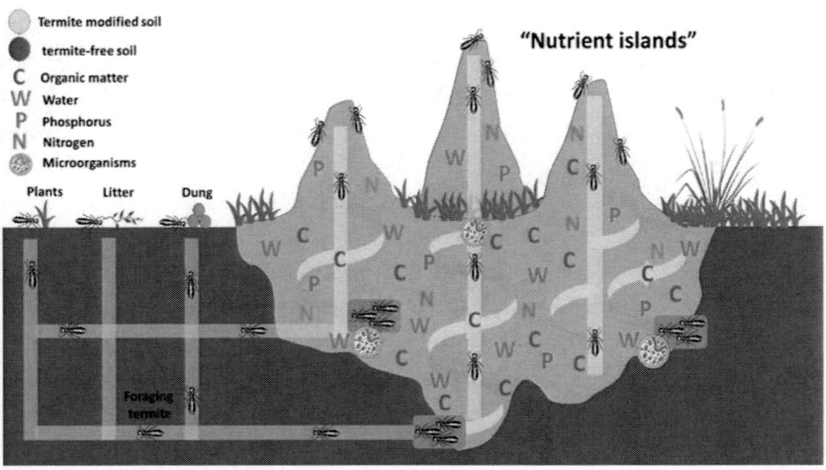

Source: Adapted from Coventry et al., (1988).

Figure 7. Framework of general termites' communities and their "nutrient island" effect on soil properties as organic matter and nutrients flow.

Micromorphological analysis of soils shows that termites, as faecal or chewing animals, develop small ovoid or spherical structures as pellets (50-1000μm) of organic soil material in their mounds, galleries or nests (Abe et al., 2000). These structures have been studied through the interaction with the development of fungal microorganisms known as fungus-growing termites (FGT) (Jones et al., 1994). Indeed, FGTs are known for their enormous capacity to mineralise organic carbon in arid savannas (Aanen and Eggleton, 2005). Furthermore, both foraging, soil-feeding and fungus-growing termites are capable of altering the formation and stability of soil aggregates. For example, soil-feeding termites were found to increase the stability of soil aggregates > 2 mm, whereas FGT reduced it through the translocation process and non-incorporation of faeces to the detriment of the use of saliva (Contour-Ansel et al., 2000; Jouquet et al., 2011).

Researchers have attempted to demonstrate termite bioturbation in the formation of biogenic sand-sized aggregates (called pseudosand aggregates) and granular organisation in tropical soils (Millogo et al., 2011; Reatto et al., 2009). However, according to Six et al., (2004), more evidence is needed on the actual impact of termites on macro- and microaggregate formation. Jouquet et al., (2022) evaluated factors controlling particle size fractions, structural stability and chemical properties of soil films produced by termite species from 21 countries, covering all feeding and building categories. They showed that environmental characteristics, in particular the bulk density of the soil, can explain the properties of termite sheeting. The classical hypothesis is that the type of feeding and building strategies of termites (FGT or non-fungus growing species) could differentiate the effects on carbon content and structural stability. They suggest that there is a need to redefine termite functional groups based on morphological, anatomical and/or physiological traits. Therefore, future studies need to take these observations into account to understand the true effect of termites on different soil processes, such as aggregation.

Soil Organic Matter in the Formation and Stabilization of Soil Aggregates

Soil organic matter (SOM) corresponds to a complex and heterogeneous mixture of organic compounds in different stages of decomposition (Cotrufo et al., 2015; Kogel-Knabner, 2002). It is mainly formed by the continuous deposition of above- and below-ground plant residues (root exudates and root tissue renewal) on the soil, as well as by-products of microbial metabolism (Cotrufo and Lavallee, 20-22; Kogel-Knabner, 2002). As a binding agent for soil particles, SOM is classified as: temporary (roots and fungal hyphae), transient (mainly by polysaccharides) or persistent (humified MO and polyvalent metal cation complexes) (Tisdall and Oades, 1982). Because of these properties, SOM provides many benefits to the soil. One of its functions is closely related to soil aggregation. These organic agents are mainly mediated by the activity of microorganisms that decompose the organic matter present in the soil, releasing mucilage and microbial metabolites (Golchin et al., 1994). In addition, microorganisms produce extracellular polymeric substances (EPS), which can increase the adhesion of cells to mineral surfaces (Kleber et al., 2015), favoring the aggregation process. EPS are mainly composed of polysaccharides and proteins, and to a lesser extent nucleic acids and lipids (Badireddy et al., 2008; Hoffman and Decho, 1999; Kleber et al., 2015; Omoike and Chorover, 2004). Roots are also important for aggregation as they produce root exudates that act as binding agents for soil particles.

Within the microbial community, fungi and bacteria are the most important organisms for soil aggregation (Brady and Weil, 2013). The most important fungi in the aggregation process are those that associate with plant roots (mycorrhizal fungi) because they secrete proteins (glomalins) that are efficient cementing agents, allowing sand and silt particles to bond together (fine aggregation) (Brady and Weil, 2013). On the other hand, bacteria release polysaccharides and other types of organic matter that are also important for particle cohesion (Brady and Weil, 2013).

The role of SOM in the formation and stabilization of soil aggregates begins with the action of free organic matter (fine roots and fungal hyphae) in binding individual soil particles and small microaggregates into larger aggregates, the macroaggregates (Brady and Weil, 2013; Six et al., 2004; Tisdall and Oades, 1982). This binding occurs both through the physical engagement of the particles (Kleber et al., 2015) and through the exudation of polysaccharides and other organic compounds by the roots and hyphae, which act as sticky networks (Brady and Weil, 2013). Furthermore, because they are temporary binding agents (Tisdall and Oades, 1982), roots and hyphae become particulate organic matter (POM) after their death and decomposition by microorganisms (Bucka et al., 2021). During this process, mucilage production and proliferation of saprotrophic hyphae occur, which also promote aggregation by encrusting clay particles to these organic compounds (Jastrow and Miller, 1998). As previously discussed, this process results in occluded microaggregates within macroaggregates (Oades, 1984) (Figure 8).

In microaggregate formation, MOS is responsible for binding mineral particles, mainly silt and clay (Brady and Weil, 2013), to organic compounds in highly stable organometallic associations (Wagai et al., 2020). The interactions between the mineral fraction and the organic compounds (mainly carboxylic and alcoholic/phenolic functional groups) can occur by adsorption (specific or non-specific) and/or by hydrogen bonds, van der Waals interactions, hydrophobic interactions, co-precipitation of organic complexes - metal, ligand exchange, anion exchange and cation bridge (Kleber et al., 2015; Sarkar et al., 2018; Stotzky, 1986), all with different intensities (Kleber et al., 2015). These interactions are influenced by a number of factors that may or may not be related to soil solution, such as pH, solution composition, binder structure, solution concentration and residence time (Kleber et al., 2015). The intensity with which these factors affect organometallic interactions will control the formation and stability of soil aggregates.

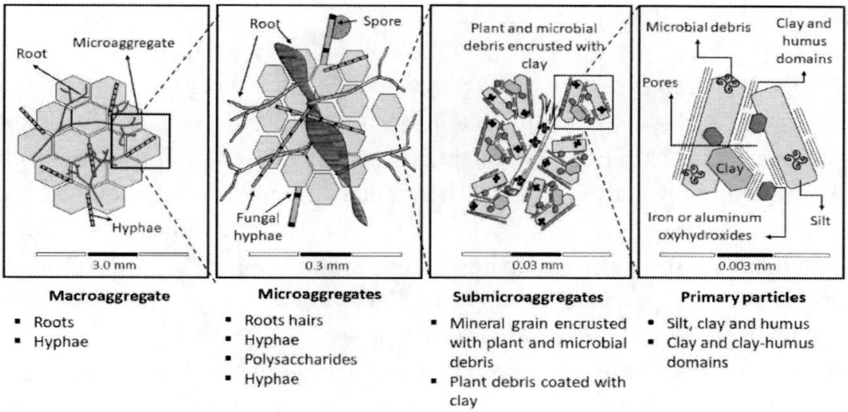

Figure 8. Scheme of the aggregation process of primary soil particles mediated by SOM and microorganisms. Adapted from Brady and Weil (2010).

Indicators for Assessing Soil Structural Stability

The laboratory methods most commonly used to evaluate aggregate stability can be grouped into (i) methods that evaluate the amount of clay dispersed in the soil (WDC); (ii) methods that evaluate the distribution of aggregates in different size classes or by general indices, as in the case of mean weight diameter (MWD) or geometric mean diameter (GMD); and (iii) the aggregate tensile strength (TS) method after applying an aggregate breaking strength.

Clay dispersion is related to the strength of the soil structure and is of agricultural and environmental importance. Dispersed clay can be transported and accumulate in deeper soil layers. Pore clogging caused by clay accumulation can reduce soil water infiltration rates and trigger erosion processes (Dexter et al., 2011). The more flocculated the soil, the more stable it is in water, and the reverse is true if the soil is highly dispersed. The chemical composition of the soil affects the flocculation/dispersion processes, as discussed above, because it influences the properties of the DDL. As an example, it can be seen that soils with high Na^+ content are more dispersive than soils with high Ca^{2+} content (Czyz and Dexter, 2015). Currently, some studies have used two measures of water-dispersible clay that are considered more sensitive (Batista et al., 2020; Nunes et al., 2021). When a known mass of soil (usually 4-5 g) is placed in water and subjected to a small standard input of mechanical energy, it is called readily dispersible clay (RDC), while the fraction of clay released from the soil without mechanical energy input is called spontaneously dispersible clay (SDC) (Dexter et al., 2011; Czyz and Dexter, 2015). Batista et al., (2020) used the readily dispersed clay (RDC) and mechanically dispersed clay (MDC) indices by the turbidimetry method to evaluate the structural stability of six different soil cover cropping systems for rhizospheric soil and bulk soil. The evaluated soil naturally showed high structural stability as it is a very clayey Oxisol. The rhizospheric soil showed higher structural stability compared to the bulk soil (Figure 9 a, b). Among the indices used, MDC was the only one that showed a significant interaction (Figure 9b). The more diversified cropping system (iCL) was the most recommended because it showed greater structural stability and accumulation of soil organic carbon (SOC), an essential factor for increasing aggregate stability.

Aggregate stability analysis by wet and dry sieving is traditionally used in soil physics. The method was first proposed by Yoder (1936) and over time has undergone several modifications and adaptations (Kemper, 1965; Kemper and Rosenau, 1986). In this case, the stability of the aggregates is measured by their distribution on sieves of different sizes and also by the DMP and DMG indices. The higher the DMP, the greater the mass of aggregates on larger sieve sizes and, therefore, the high structural stability of a given soil (Kemper and Rosenau, 1986). However, some limitations of the method may be presented in tropical and/or highly weathered soils. This can occur because these soils often have a highly stable microgranular structure. This structure is influenced by the presence of iron and aluminum oxides (Ferreira et al., 1999; Pessoa et al., 2022). Thus, a larger mass of aggregates present on smaller sieve sizes for these soils will not always indicate low structural stability, due to the frequency of microaggregates (< 250 μm).

Tensile strength (TS) is related to integrating processes leading to strengthening or weakening of fault zones (Kay and Dexter, 1992). Individual soil aggregates can be measured using TS testing. TS measures the force required to break the aggregate in the fracture planes and is considered a sensitive indicator of aggregate stability; TS of spherical particles could be determined from a simple crushing test (Dexter and Kroesbergen, 1985).

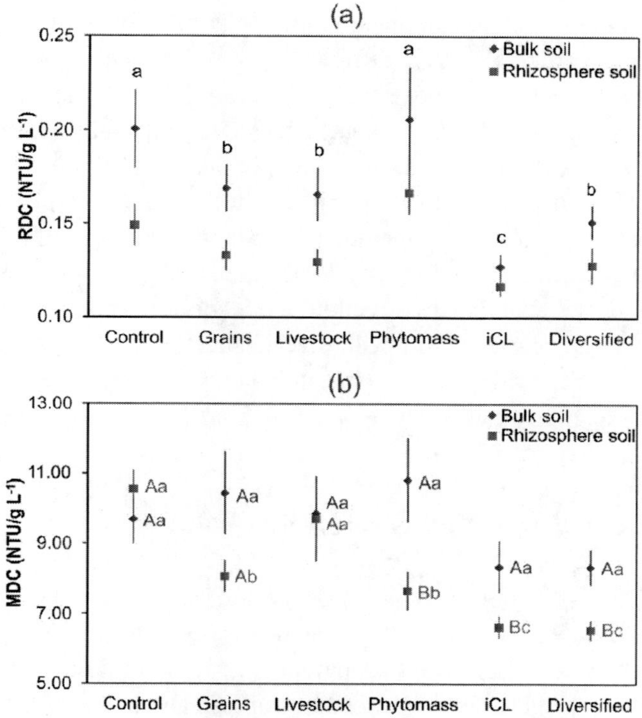

Source: Batista et al., (2020).

Figure 9. (a) RDC and (b) MDC in surface soil layer (0–0.05 m) of rhizosphere soil and bulk soil. Different letters represent significant differences by Scott Knott's test at $\alpha = 0.05$. For RDC (a) letters represent differences among treatments, and for MDC (b) uppercase letters represent differences between environments (rhizosphere soil or bulk soil) for each treatment, and the lowercase letters represent differences between the treatments for each environment.

TS can be easily defined as the ability of a soil body to withstand forces without failure, the maximum stress that a given aggregate can withstand without failure, or simply the minimum stress to failure of a body. However, it is worth noting that TS is a variable that exhibits high variability, can change during the measurement process, and the deformed soil body tends to increase or decrease its strength as new deformations occur (Hillel, 2003).

Felde et al., (2021) developed a new dry-crushing method for isolating microaggregates along rupture planes under mechanical pressures. The method was compared to traditional wet sieving. The results showed that dry crushed and sieved microaggregates presented higher aggregate stability. Dry microaggregates were larger in size and contained more sand particles. On the other hand, in the wet sieving method, the aggregates showed more clay-size and silt-size particles. Thus, differences in aggregate stability could be observed based on the method used. Obalum et al., (2019) evaluated aggregate stability as a function of management effects. The soil analyzed was a Luvisol silt loam in reduced tillage and conventional tillage in Central Germany about 2.5 years after the start of the experiment. In the study, 12 different indices from different laboratory methods were used. Four of the indices discriminated between the tillage systems; three (wet sieving - MWDw, water stable aggregates with sand correction - WSAcfs and tensile strength of re-moistened aggregates - TSrm with size between 16-8 mm showed higher stability under reduced tillage), on the other hand, the aggregate stability index (ASI) showed an opposite effect.

Soil Structure as an Indicator of Soil Quality

Soil structure is a physical indicator of soil quality and is, directly and indirectly, directly and indirectly, related to several other variables and soil functional processes (Lal, 2004; Dantas et al., 2012). In addition, soil structure is positively or negatively influenced by soil use and management (Cherubin et al., 2017; Silva et al., 2018; Falcão et al., 2020). Therefore, soil structure is a key indicator for assessing soil quality and health. Ultimately, soil structure is a variable or aspect of soil that allows the assessment of soil functioning.

Concerned about soil degradation around the world, soil scientists have broadened and deepened the discussion of soil quality and sought to understand its relationship to the sustainability of agricultural ecosystems. The first concepts of soil quality were introduced in the 1990s (Doran; Parkin, 1994; Johnson et al., 1997). The concept of soil quality (SQ) can be briefly defined as the ability of a given soil to function within the limits of the ecosystem in which it is located. Doran and Parkin (1994) defined SQ as the capacity of soil to support the production of plants and animals and to maintain or improve the quality of air and water, thereby ensuring the health of humans, plants, and animals, within the limits of the natural or managed ecosystem in which it is located. Johnson et al., (1997) defined soil quality as “a measure of soil condition in relation to the requirements of one or more biological species and/or for some human purpose”. Due to the importance of this issue, several studies have been carried out worldwide over the last 30 years to investigate the effects of soil management systems on soil quality. Soil quality has also gained importance for soils in tropical climates and several studies have been carried out in the last decades (Araújo; Goedert; Lacerda, 2007; Cardoso et al., 2009; Cherubin et al., 2015, 2017; Lisboa et al., 2012; Marinho et al., 2021; Sales et al., 2018).

Linking agricultural and soil science with policy, stakeholder needs, and sustainable supply chain management, soil health is the continued ability of soil to function as a vital living ecosystem that supports plants, animals, and humans at global or even at field Pedon scales (Lehmann et al., 2020). From this concept, the degradation of soil structural quality at a local scale affects crop production through reduced porosity and increased soil density, reduced capacity of the soil to infiltrate, store and make available water, and increased resistance of the soil to root infiltration. On a global scale, soil structural degradation affects the soil's ability to store carbon, cycle nutrients, supply underground water reservoirs and ultimately maintain or improve the health of the crops, animal, and human populations.

It is important to make a few comments about these SQ concepts. First, SQ is directly linked to an objective, which can vary from the production of food, fiber, energy, carbon sequestration, and aquifer recharge, among others. Second is the concept's scope, as it contextualizes soil functioning within the boundaries of the immediate ecosystem in which the soil is embedded. In this respect, it is reasonable to extend the understanding to managed ecosystems, which may have different levels of complexity. This relationship will be discussed later in this theme.

Despite the efforts of soil scientists around the world, it is not possible to measure or assess SQ directly. To quantify and qualify SQ, it is necessary to analyze a set of soil variables or attributes called soil quality indicator attributes (Bünemann et al., 2018; Lal, 2016; Tótola and Chaer, 2002; Vezzani and Mielniczuk, 2009). The use of variables or attributes to measure and assess SQ is based on the premise that there is a relationship between soil attributes and processes occurring in the soil, and therefore a relationship between soil attributes and soil functioning. In addition to being related to soil processes, it is necessary for the attribute to be sensitive to management impacts on the soil, easy and inexpensive to measure, understandable, and related to the processes of the environment in which the soil is situated. Because they are related to environmental conditions, soil attributes can be indicators of the state of soil degradation (Martins et al., 2010).

Conceptually, soil quality indicator attributes are grouped into three categories: chemical, physical and biological indicators. Despite this conceptual categorization, the indicators interact with each other in the soil and the result of the interaction of these attributes leads to soil functioning. It is impossible to infer soil quality from a single attribute (indicator). Therefore, a set of variables is usually analyzed and combined using bivariate and/or multivariate techniques to infer soil quality (Dantas et al., 2012; De Morais et al., 2015; Maikhuri et al., 2012). Similarly, it is not possible to establish a universal index for assessing soil quality. This is due to the great diversity of soils and ecosystems in the world, the "limited occurrence of soils" and the fact that soil quality is conceptually linked to the objective of ecosystem management. Soil variables such as pH, nutrient content, available content of potentially toxic elements, carbon content, soil structure, soil density, soil porosity, activity, and microbial diversity are commonly used as attributes indicating soil quality (Araújo et al., 2012; Araújo et al., 2007; Bünemann et al., 2018; Capuani et al., 2012).

Some studies propose variables related to crop production as an indicator of soil quality, as biomass production is largely influenced by soil conditions (Cui et al., 2021; Lal, 2004). The soil production function has long been recognized by man, and perhaps for this reason it is almost intuitive to think of productivity as a soil quality index (SQI). In general, biological quality indicators are the most sensitive to land use changes (Carneiro et al., 2009; Lehman et al., 2020; Marinho Júnior et al., 2021; Martins et al., 2010; Vezzani and Mielniczuk, 2009).

Although there is no consensus on a standard methodology for assessing SQ, researchers have attempted to develop an SQI based on soil attributes. The SQI is a methodology for assessing soil quality in a way that aggregates the contribution of several soil quality indicator attributes. In practice, the soil quality index is used as a decision-making tool oriented toward ecosystem management with the aim of achieving sustainability (Bünemann et al., 2018; Karaca et al., 2021; Lehmann et al., 2020).

In brief, the SQI is developed in four basic steps: 1 - the indicators are measured in the field and/or soil samples are collected for laboratory analysis; 2 - the indicators that will make up the SQI are selected, empirically or by applying statistical techniques (this step is commonly referred to as obtaining the minimum set of data); 3 - the data obtained from the measurement of the indicators are standardized; 4 - the selected indicators are combined according to the methodology for obtaining the index. The SQI values obtained for the farmed soils are compared with the SQI values observed for the soils in the reference areas.

An important practical implication of the impossibility of directly measuring and assessing SQ is the need to establish reference areas. In general, areas of native vegetation under the same edaphoclimatic conditions with little or no anthropic influence are established as reference areas. It is assumed that in native areas the soil functions adequately, taking into account the limits of the ecosystem.

Effects such as reduction in soil C content (total carbon, microbial carbon) (Cardoso et al., 2009; Da silva et al., 2009), changes in soil structure (increase in density and decrease in soil porosity) (Aratani et al., 2009; Araújo et al., 2007; Oliveira et al., 2015) and effects on soil biological properties (reduction of enzymatic activity) (Da silva et al., 2009) are often observed in managed soils compared to soils from natural ecosystems.

Soil management by tillage systems increases the turnover of soil aggregates and consequently, the organic matter sequestered by land use conversion from native ecosystem to agriculture (Six et al., 2000). It has also been observed that no-tillage (NT) systems lead to an accumulation of soil organic matter by reducing soil disturbance compared to conventional tillage (CT) (Oades, 1984). This is due to the increased input of organic matter, the redistribution of residues in aggregates (protection), and the reduction of aggregate turnover (Oades, 1984; Paustian et al., 2000). Six et al. (2000) investigated the relationship between soil aggregation turnover and soil organic matter in the NT system. They observed that the rate of macroaggregate formation and degradation is reduced under NT compared to CT. It was also observed that soil aggregation turnover under NT leads to the formation of stable microaggregates in which carbon is stabilized.

Soil tillage and heavy machine traffic can cause soil compaction, consequently, all disturbance decreases the aggregate stability. Additionally, soil fertilization can decrease aggregate stability by altering the loads and elements present in the soil (Nunes et al., 2017). Anthonio et al. (2022) studied the effects of the combination of traditional fertilizers such as NPK and pig manure on the stability of aggregates in a rice-managed paddy located in Nanchang County, China since 1984. The treatments consisted of unfertilized control (CK), only NPK, 30% NPK + 70% pig manure (30F), 50% NPK + 50% pig manure (50F), and 70% NPK + 30% pig manure (70F) in a Ferralic Cambisol with a heavy clayey texture. The main results showed that treatment with NPK and pig manure increased aggregate stability. The pig manure treatment showed the highest proportions of macroaggregates compared to other treatments. The pig manure favored the increase of mean weight diameter (MWD) and geometric mean diameter (GMD). The MWD ranged from 1.04 mm (30F) to 0.81 mm (CK) and 0.94 mm (30F) to 0.72 mm (CK) at the surface and subsurface, respectively. Therefore, the importance of soil organic matter (SOM) in increasing soil structural stability was evident.

In contrast, SOM did not increase aggregate stability in an organic farming area. A study assessed the change provided in the soil in an organic farming area over seven years on a Vertisol in India, with a two-year crop rotation of cotton and soybean-wheat. The area received ~13.6 t ha^{-1} of composted manure annually, compared to a conventionally farmed area, which received only 2 t ha^{-1} of poultry manure + ~ 170 kg N ha^{-1}. Even after a high application of compost organic, the organic carbon (C_{org}) did not present differences. The results showed that the application of organic correctives may not increase C_{org} contents and improve the soil's structural stability (Schweizer et al., 2022). These observations show that even soil aggregates are affected. Soil aggregates, even on a very small-scale size, are important components for the carbon cycle and sequestration.

Conclusion

Soil aggregation is now, more than ever, being observed as one of the soil properties capable of revealing soil quality. This ability is due to the fact that the aggregate microenvironment can gather soil forming and maintaining factors such as soil organic matter and carbon stocks, soil fauna effects, and microbiology, such as binding agents, in addition to the ability of aggregates to resist different management pressures. Soil use and management practices are expected to evolve towards more integrated and conserving systems that preserve soil structure and, consequently, the biogeochemical factors involved in maintaining it.

References

Aanen, D. K., Eggleton, P., 2005. Fungus-growing termites originated in African rain forest. *Current Biology* 15: 851–855.

Abe, T., Bignell, D. E., Higashi, M., Higashi, T., Abe, Y. *Termites: evolution, sociality, symbioses, ecology*. Springer Science & Business Media, 2000.

Afriyie, E., Verdoodt, A., Mouazen, A. M., 2022. Potential of visible-near infrared spectroscopy for the determination of three soil aggregate stability indices. *Soil and Tillage Research* 215: 105218.

Alleoni, L. R. F., Mello, J. W. V. D., Rocha, W. S. D. D. Eletroquímica, adsorção e troca iônica no solo. [Electrochemistry, adsorption and ion exchange in soil.] In: Melo, V. F., Alleoni, L. R. F (Eds). *Química e mineralogia do solo – Parte II [Soil chemistry and mineralogy – Part II]*. Viçosa: SBCS, 2009.

Amelung, W., Meyer, N., Rodionov, A., Knief, C., Aehnelt, M., Bauke, S. L., Biesgen, D., Dultz, S., Guggenberger, G., Jaber, M., Klumpp, E., Kögel-Knabner, I., Nischwitz, V., Schweizer, S. A., Wu, B., Totsche, K. U., Lehndorff, E., 2023. Process sequence of soil aggregate formation disentangled through multi-isotope labelling. *Geoderma* 429: 116226.

Angers, D. A., Recous, S., Aita, C., 1997. Fate of carbon and nitrogen in water-stable aggregates during decomposition of $_{13}C_{15}N$-labelled wheat straw in situ. *European Journal of Soil Science* 48: 295–300.

Anthonio, C. K., Huang, J., Han, T., Qaswar, M., Liu, K., Du, J., Daba, N. A., Ali, S., Matele, L. A., Liu, S., Liu, G., Zhang, H., Chen, J., 2022. Fertilizer combination effects on aggregate stability and distribution of aluminum and iron oxides. *Journal of Plant Nutrition and Soil Science* 185:251-263.

Aratani, R. G., Freddi, O. S., Centurion, J. F., Andrioli, I., 2009. Qualidade física de um Latossolo Vermelho acriférrico sob diferentes sistemas de uso e manejo [Physical quality of an acryferric Red Latosol under different systems of use and management]. *Revista Brasileira de Ciência do Solo [Brazilian Journal of Soil Science]* 33: 677–687.

Araújo, E. A., Ker, J. C., Neves, J. C. L., Lani, J. L., 2012. Qualidade do solo: conceitos, indicadores e avaliação [Soil quality: concepts, indicators and evaluation]. *Pesquisa Aplicada & Agrotecnologia [Applied Research & Agrotechnology]* 5:187–196.

Araújo, R., Goedert, W. J., Lacerda, M. P. C., 2007. Qualidade de um solo sob diferentes usos e sob Cerrado nativo [Quality of a soil under different uses and under native Cerrado]. *Revista Brasileira de Ciência do Solo [Brazilian Journal of Soil Science]* 31:1099–1108.

Babu Ojha, R., Devkota, D., 2014. Earthworms: "Soil and Ecosystem Engineers" – a Review. *World Journal of Agricultural Research* 2: 257–260.

Badireddy, A. R., Korpol, B. R., Chellam, S., Gassman, P. L., Engelhard, M. H., Lea, A. S., Rosso, K. M., 2008. Spectroscopic characterization of extracellular polymeric substances from Escherichia coli and Serratia marcescens: Suppression using sub-inhibitory concentrations of bismuth thiols. *Biomacromolecules 9*(11): 3079–3089.

Batista, A. M., Libardi, P. L., Giarola, N. F. B., 2020. Evaluation of the soil aggregation induced by the plant roots in an Oxisol by turbidimetry and water percolation. *Rhizosphere* 16: 100265.

Bossuyt, H., Six, J., Hendrix, P. F., 2004. Rapid incorporation of carbon from fresh residues into newly formed stable microaggregates within earthworm casts. *European Journal of Soil Science* 55: 393–399.

Bossuyt, H., Six, J., Hendrix, P. F., 2005. Protection of soil carbon by microaggregates within earthworm casts. *Soil Biology and Biochemistry* 37: 251–258.

Bouché, M. B. The establishment of earthworm communities. In: Satchell, J. E. (Ed.), *Earthworm Ecology: From Darwin to Vermiculture.* Dordrecht: Springer Netherlands, 1983.

Brady, N. C., Weil, R. R. Elementos da Natureza e Propriedades dos Solos [Elements of Nature and Properties of Soils]. In: Lepsch, I. F. (Ed.), *Elementos da Natureza e Propriedades dos Solos [Elements of Nature and Properties of Soils]*. Bookman, 2013.

Brown, G. G., Barois, I., Lavelle, P., 2000. Regulation of soil organic matter dynamics and microbial activity in the drilosphere and the role of interactions with other edaphic functional domains. *European Journal of Soil Biology* 36: 177–198.

Bucka, F. B., Felde, V. J. M. N. L., Peth, S., Kögel-Knabner, I., 2021. Disentangling the effects of OM quality and soil texture on microbially mediated structure formation in artificial model soils. *Geoderma* 403:115213.

Bünemann, E. K., Bongiorno, G., Bai, Z., Creamer, R. E., De Deyn, G., de Goede, R., Fleskens, L., Geissen, V., Kuyper, T. W., Mäder, P., Pulleman, M., Sukkel, W., van Groenigen, J. W., Brussaard, L., 2018. Soil quality – A critical review. *Soil Biology and Biochemistry* 120: 105–125.

Capuani, S., Rigon, J. P., Beltrão, N. E. D. M., Brito Neto, J. F. D., 2012. Atividade microbiana em solos, influenciada por resíduos de algodão e torta de mamona [Microbial activity in soils, influenced by cotton residues and castor bean cake]. *Revista Brasileira de Engenharia Agrícola e Ambiental [Brazilian Journal of Agricultural and Environmental Engineering]* 16:1269–1274.

Cardoso, E. L., Silva, M. L. N., Moreira, F. M. D. S., Curi, N., 2009. Atributos biológicos indicadores da qualidade do solo em pastagem cultivada e nativa no Pantanal [Biological attributes indicators of soil quality in cultivated and native pasture in the Pantanal]. *Pesquisa Agropecuária Brasileira [Brazilian Agricultural Research]* 44:631–637.

Carneiro, M. A. C., Souza, E. D. D., Reis, E. F. D., Pereira, H. S., Azevedo, W. R. D., 2009. Atributos físicos, químicos e biológicos de solo de cerrado sob diferentes sistemas de uso e manejo. [Physical, chemical and biological attributes of Cerrado soil under different systems of use and management.] *Revista Brasileira de Ciência do Solo [Brazilian Journal of Soil Science]* 33:147–157.

Cherubin, M. R., Eitelwein, M. T., Fabbris, C., Weirich, S. W., Silva, R. F. D., Silva, V. R. D., Basso, C. J., 2015. Physical, chemical, and biological quality in an Oxisol under different tillage and fertilizer sources. *Revista Brasileira de Ciência do Solo [Brazilian Journal of Soil Science]* 39: 615–625.

Cherubin, M. R., Franco, A. L., Guimarães, R. M., Tormena, C. A., Cerri, C. E., Karlen, D. L., Cerri, C. C., 2017. Assessing soil structural quality under Brazilian sugarcane expansion areas using Visual Evaluation of Soil Structure (VESS). *Soil and Tillage Research* 173: 64–74.

Contour-Ansel, D., Garnier-Sillam, E., Lachaux, M., Croci, V., 2000. High performance liquid chromatography studies on the polysaccharides in the walls of the mounds of two species of termite in Senegal, Cubitermes oculatus and Macrotermes subhyalinus: their origin and contribution to structural stability. *Biology and Fertility of Soils* 31: 508–516.

Cotrufo, M. F., Lavallee, J. M. Soil organic matter formation, persistence, and functioning: A synthesis of current understanding to inform its conservation and regeneration. In: Donald. L. Sparks (Ed.). *Advances in Agronomy*, 2022.

Cotrufo, M. F., Soong, J. L., Horton, A. J., Campbell, E. E., Haddix, M. L., Wall, D. H., Parton, W. J. 2015. Formation of soil organic matter via biochemical and physical pathways of litter mass loss. *Nature Geoscience* *8*(10): 776–779.

Coventry, R. J., Holt, J. A., Sinclair, D. F., 1988. Nutrient cycling by mound building termites in low fertility soils of semi-arid tropical Australia. *Soil Research* 26: 375–390.

Cui, Q., Xia, J., Yang, H., Liu, J., Shao, P., 2021. Biochar and effective microorganisms promote Sesbania cannabina growth and soil quality in the coastal saline-alkali soil of the Yellow River Delta, China. *Science of The Total Environment* 756: 143801.

Czyz, E. A., Dexter, A. R., 2015. Mechanical dispersion of clay from soil into water: readily-dispersed and spontaneously-dispersed clay. *International Agrophysics* 29: 1-7.

Dantas, J. D. A. N., Oliveira, T. S., Mendonca, E. S., Assis, C. P., 2012. Soil quality under different land use and management in the Irrigated Perimeter of Jaguaribe/Apodi, Brazil. *Revista Brasileira de Engenharia Agrícola e Ambiental [Brazilian Journal of Agricultural and Environmental Engineering]* 16:18–26.

De Bruyn, L. A. L., Conacher, A. J., 1990. The role of termites and ants in soil modification-a review. *Soil Research* 28: 55–93.

de Morais, E. R., Maia, C. E., Gaudêncio, H. R., Sousa, D. M., 2015. Indices of chemical quality of soil in areas cultivated with irrigated papaya. *Revista Brasileira de Engenharia Agrícola e Ambiental [Brazilian Journal of Agricultural and Environmental Engineering]* 19:587–591.

Dexter, A. R., Kroesbergen, B., 1985. Methodology for determination of tensile strength of soil aggregates. *Journal of Agricultural Engineering Research* 31: 139-147.

Dexter, A. R., Richard, G., Czyz, E. A., Davy, J., Hardy, M., Duval, O., 2011. Clay Dispersion from Soil as a Function of Antecedent Water Potential. *Soil Science Society of America Journal* 75: 444-455.

Doran, J. W., Parkin, T. B. Defining and assessing soil quality. In: *Defining soil quality for a sustainable environment*. Madison: Soil Science Society of America, 1994.

Doran, J. W., Parkin, T. B., 1996. Quantitative indicators of soil quality: a minimum data set. *Methods for Assessing Soil Quality* 49: 25–37.

Edwards, A. P., Bremner, J. M., 1967. Microaggregates in Soils. *European Journal of Soil Science* 18: 64-73.

Edwards, C. A., Bohlen, P. J. *Biology and ecology of Earthworms*. Springer Science & Business Media, 1996.

Falcão, K. D. S., Monteiro, F. D. N., Ozório, J. M. B., Souza, C. B da S., Farias, P. G da S., Menezes, R. D. S., Panachuki, E., Rosset, J. S., 2020. Estoque de carbono e agregação do solo sob diferentes sistemas de uso no cerrado [Carbon stock and soil aggregation under different use systems in the cerrado]. *Revista Brasileira de Ciências Ambientais [Brazilian Journal of Environmental Sciences]* 55:242-55.

Felde, V. J. M. N. L., Schweizer, S. A., Biesgen, D., Ulbrich, A., Uteau, D., Knief, C., Graf-Rosenfellner. M., Kögel-Knabner, I., Peth, S., 2021. Wet sieving versus dry crushing: Soil microaggregates reveal different physical structure, bacterial diversity and organic matter composition in a clay gradient. *European Journal of Soil Science* 72: 810-828.

Ferreira, M. M., Fernandes, B., Curi, N., 1999. Mineralogia da fração argila e estrutura de Latossolos da região sudeste do Brasil. [Mineralogy of the clay fraction and structure of Oxisols from southeastern Brazil.] *Revista Brasileira de Ciência do Solo [Brazilian Journal of Soil Science]* 23: 507-514.

Golchin, A., Oades, J. M., Skjemstad, J. O., Clarke, P. 1994. Study of Free and Occluded Particulate Organic Matter in Soils by Solid state 13c CP/MAS NMR Spectroscopy and Scanning Electron Microscopy. *Aust. J. Soil Res* 32: 285–309.

Golchin, A., Oades, J. M., Skjemstad, J. O., Clarke, P., 1994. Soil structure and carbon cycling. *Soil Research* 32: 1043–1068.

Guhra, T., Stolze, K., Schweizer, S., Totsche, K. U., 2020. Earthworm mucus contributes to the formation of organo-mineral associations in soil. *Soil Biology and Biochemistry* 145: 107785.

Guhra, T., Stolze, K., Totsche, K. U., 2021. Pathways of biogenically excreted organic matter into soil aggregates. *Soil Biology and Biochemistry* 164: 108483.

Hallam, J., Hodson, M. E., 2020. Impact of different earthworm ecotypes on water stable aggregates and soil water holding capacity. *Biology and Fertility of Soils* 56: 607–617.

Harit, A., Shanbhag, R., Chaudhary, E., Cheik, S., Jouquet, P., 2017. Properties and functional impact of termite sheetings. *Biology and Fertility of Soils* 53: 743–749.

Hoffman·, M., Decho, A. W. Extracellular Enzymes Within Microbial Biofilms and the Role of the Extracellular Polymer Matrix. In: Wingender, J., Neu, T. R., Flemming, H. C. (Eds.). *Microbial Extracellular Polymeric Substances*. Berlin Heidelberg: Springer, 1999.

Holt, J. A., Lepage, M., 2000. Termites and soil properties. Termit. Evol. Soc. symbioses, *Ecol.* 389–407.

Jastrow, J. D., Miller, R. M. Soil Aggregate Stabilization and Carbon Sequestration: Feedbacks through Organomineral Associations. In: Lal, R., Kimble, J. M., Follett, R. F., Stewart, B. A. (Eds.). *Soil Processes and the Carbon Cycle* CRC Press, 1998.

Johnson, D. L., Ambrose, S. H., Bassett, T. J., Bowen, M. L., Crummey, D. E., Isaacson, J. S. & Winter-Nelson, A. E., 1997. Meanings of Environmental Terms. *Journal of Environmental Quality* 26:581–589.

Jones, C. G., Lawton, J. H., Shachak, M., 1994. Organisms as ecosystem engineers, in: *Ecosystem Management.* Springer, pp. 130–147.

Jouquet, P., Harit A., Hervé V.,Moger H., Carrijo T., Donoso D. A., Eldridge D., Cunha H. F., Choosai C., Janeau J. L., Maeght J. L., Thu T. H., Briandon A., Skali M. D., Thuyne J. V., Mainga Ali, Florian O. P. P., Issa O. M., Podwojewski P., Rajot J. L., Tureaux T. H., Smaili L., Labiadh M., Boukbida H. A., Shanbhag R., Muon R., Ann V., Cheik V., Fall S., Traoré S., Dupont S., Chouvenc T., Mullins A. J., Syaukani S., Zaiss R., Tien T. M., Šobotník J., Auclerc A., Qiu R., Tang Y. T., Huot H., Sillam-Dussès D., Bottinelli N.,2022. The impact of termites on soil sheeting properties is better explained by environmental factors than by their feeding and building strategies. *Geoderma* 412: 115706.

Jouquet, P., Traoré, S., Choosai, C., Hartmann, C., Bignell, D., 2011. Influence of termites on ecosystem functioning. Ecosystem services provided by termites. *European Journal of Soil Biology* 47: 215–222.

Karaca, S., Dengiz, O., Turan, I. D., Özkan, B., Dedeoğlu, M., Gülser, F, Sargin, B., Demirkaya, S., Ay, A., 2021. An assessment of pasture soils quality based on multi-indicator weighting approaches in semi-arid ecosystem. *Ecological Indicators* 121:107001.

Kay, B. D., Dexter, A. R., 1992. The influence of dispersible clay and wetting/drying cycles on the tensile strength of a red-brown earth. *Soil Research* 30: 297-310.

Kemper, D. W., Rosenau, R. C. Aggregate stability and size distribution. In: Klute, A. (Ed.). *Methods of Soil Analysis Part 1*. Madison: American Society of Agronomy, 1986.

Kleber, M., Eusterhues, K., Keiluweit, M., Mikutta, C., Mikutta, R., Nico, P. S., 2015. Mineral-Organic Associations: Formation, Properties, and Relevance in Soil Environments. *Advances in Agronomy* 130:1–140.

Kogel-Knabner, I., 2002. The macromolecular organic composition of plant and microbial residues as inputs to soil organic matter. *Soil Biology e Biochemistry* 34: 139–162.

Lal, R. *Soil health and carbon management. Food and Energy Security* Wiley-Blackwell Publishing, 2016.

Lal, R., 2004. Soil Carbon Sequestration Impacts on Global Climate Change and Food Security. *Science* 304:1623–1628.

Lal, R., Shukla, M. K. *Principles of Soil Physics. CRC* Press, 2004.

Lavelle, P., Blanchart, E., Martin, A., Martin, S., Spain, A., 1993. A hierarchical model for decomposition in terrestrial ecosystems: application to soils of the humid tropics. *Biotropica* 25*:* 130–150.

Lavelle, P., Martin, A., 1992. Small-scale and large-scale effects of endogeic earthworms on soil organic matter dynamics in soils of the humid tropics. *Soil Biology and Biochemistry* 24: 1491–1498.

Lee, K. E., Foster, R. C., 1991. Soil fauna and soil structure. *Soil Research* 29: 745–775.

Lee, K. E., Wood, T. G. *Termites and soils.* London: Academic Press, 1971.

Lehmann, J., Bossio, D. A., Kögel-Knabner, I., Rillig, M. C., 2020. The concept and future prospects of soil health. *Nature Reviews Earth and Environment* 1:544–553.

Lisboa, B. B., Vargas, L. K., Silveira, A. O. D., Martins, A. F., Selbach, P. A., 2012. Indicadores microbianos de qualidade do solo em diferentes sistemas de manejo. [Microbial indicators of soil quality in different management systems.] *Revista Brasileira de Ciência do Solo [Brazilian Journal of Soil Science]* 36:33–44.

Liu, T., Chen, X., Gong, X., Lubbers, I. M., Jiang, Y., Feng, W., Li, X., Whalen, J. K., Bonkowski, M., Griffiths, B. S., Hu, F., Liu, M., 2019. Earthworms Coordinate Soil Biota to Improve Multiple Ecosystem Functions. *Current Biology* 29: 3420-3429.

Maikhuri, R. K., Kottapalli, S. R., 2012. "Soil quality and soil health: A review." *International Journal of Ecology and Environmental Sciences* 38.1: 19-37.

Marinho Junior, J. L., Piscoya, V. C., Cunha Filho, M., Fernandes, M. M., Pedrotti, A., Piscoya, T. O. F., Gomes Filho, R. R., Campos, F. S., Holanda, F. S. R., Castro, J. B., Roncal, J. L. P., Araújo Filho, R. N., 2021. Soil microbiological activity under different vegetation coverages in the Cerrado biome of Tocantins state. *Ciência Florestal [Forest Science]* 31:1494–1511.

Martin, A., 1991. Short- and long-term effects of the endogeic earthworm Millsonia anomala (Omodeo) (Megascolecidæ, Oligochæta) of tropical savannas, on soil organic matter. *Biology and Fertility of Soils* 11: 234–238.

Martin, J. P., Martin W. P., Page J. B., Raney W. A., M. D., 1955. Soil Aggregation. *Advances in Agronomy* 7: 1–37.

Martins, C. M., Galindo, I. C. D. L., Souza, E. R. D., Poroca, H. A., 2010. Atributos químicos e microbianos do solo de áreas em processo de desertificação no semiárido de Pernambuco. [Chemical and microbial attributes of the soil of areas in the process of desertification in the semiarid region of Pernambuco.] *Revista Brasileira de Ciência do Solo [Brazilian Journal of Soil Science]* 34:1883–1890.
Mccartney, D. A., Stinner, B. R., Bohlen, P. J., 1997. Organic matter dynamics in maize agroecosystems as affected by earthworm manipulations and fertility source. *Soil Biology and Biochemistry* 29: 397–400.
Millogo, Y., Hajjaji, M., Morel, J. C., 2011. Physical properties, microstructure and mineralogy of termite mound material considered as construction materials. *Applied Clay Science* 52: 160–164.
Mota, J. C. A., Freire, A. G., Alves, C. V. O., Alencar, T. L. 2017. Impactos de uso e manejo do solo na variabilidade e qualidade de atributos físicos de Cambissolos [Impacts of land use and management on the variability and quality of physical attributes of Cambisols]. *Rev agro@mbiente on-line* 11:277.
Müller-Lemans, H., van Dorp, F., 1996. Bioturbation as a mechanism for radionuclide transport in soil: Relevance of earthworms. *Journal of Environmental Radioactivity* 31: 7–20.
Nunes, M. R., Silva, A. P., Denardin, J. E., Giarola, N. F. B., Vaz, C. M. P., van Es, H. M., Silva, A. R., 2017. Soil chemical management drives structural degradation of Oxisols under a no-till cropping system. *Soil Research* 55: 819–831.
Nunes, M. R., Vaz, C. M. P., Pessoa, T. N., Denardin, J. E., Karlen, D., Giarola, N. F. B., Libardi, P. L. 2021. Inherent and dynamic effects on the structural stability of Brazilian Oxisols. *Geoderma Regional* 27: e00426.
Oades, J. M., 1984. Soil organic matter and structural stability: mechanisms and implications for management. *Plant and Soil* 76: 319–337.
Obalum, S. E., Uteau-Puschmann, D., Peth, S. 2019. Reduced tillage and compost effects on soil aggregate stability of a silt-loam Luvisol using different aggregate stability tests. *Soil and Tillage Research* 189: 217-228.
Oliveira, D. M. S., de Lima, R. P., Verburg, E. E. J., 2015. Physical quality of soil under different systems of tillage and application of pig slurry. *Revista Brasileira de Engenharia Agrícola e Ambiental [Brazilian Journal of Agricultural and Environmental Engineering]* 19:280–285.
Omoike, A., Chorover, J., 2004. Spectroscopic study of extracellular polymeric substances from Bacillus subtilis: Aqueous chemistry and adsorption effects. *Biomacromolecules* 5(4):1219–1230.
Pessoa, T. N., Cooper, M., Nunes, M. R., Uteau, D., Peth, S., Vaz, C. M. P., Libardi, P. L., 2022. 2D and 3D techniques to assess the structure and porosity of Oxisols and their correlations with other soil properties. *Catena* 210: 105899.
Plante, A. F., McGill, W. B., 2002. Soil aggregate dynamics and the retention of organic matter in laboratory-incubated soil with differing simulated tillage frequencies. *Soil and Tillage Research* 66: 79–92.
Reatto, A., Bruand, A., de Souza Martins, E., Muller, F., da Silva, E. M., de Carvalho Jr, O. A., Brossard, M., Richard, G., 2009. Development and origin of the microgranular

structure in Latosols of the Brazilian Central Plateau: significance of texture, mineralogy, and biological activity. *Catena* 76: 122–134.

Sales, A., Silva, A. R., Veloso, C. A. C., Carvalho, E. J. M., & Miranda, B. M.,2018. Carbono orgânico e atributos físicos do solo sob manejo agropecuário sustentável na Amazônia Legal [Organic carbon and soil physical attributes under sustainable agricultural management in the Legal Amazon]. *Colloquium Agrariae* 14: 01–15.

Sarkar, B., Singh, M., Mandal, S., Churchman, G. J., Bolan, N. S. Clay minerals-organic matter interactions in relation to carbon stabilization in soils. In: Garcia, C., Nannipieri, P., Hernandez, T. (Eds.), *The Future of Soil Carbon: Its Conservation and Formation.* Elsevier Inc, 2018.

Scheu, S., 1991. Mucus excretion and carbon turnover of endogeic earthworms. *Biology and Fertility of Soils* 12: 217–220.

Schweizer, S. A., Graf-Rosenfellner, M., Bhat, N. A., Kayser, G., Sisodia, B. S., Kirchhof, G., Zikeli, S., Cadisch, G., Bhullar, G. S., 2022. Responses of soil organic carbon, aggregate diameters, and hydraulic properties to long-term organic and conventional farming on a Vertisol in India. *Land Degradation & Development* 33: 785-797.

Shipitalo, M. J., Le Bayon, R. C. Quantifying the Effects of Earthworms on Soil Aggregation and Porosity. *Earthworm Ecology*. CRC Press, 2004.

Shipitalo, M. J., Protz, R., 1989. Chemistry and micromorphology of aggregation in earthworm casts. *Geoderma* 45: 357–374.

Silva, L. G. D., Mendes, I. D. C., Reis Junior, F. B., Fernandes, M. F., Melo, J. T. D., Kato, E., 2009. Atributos físicos, químicos e biológicos de um Latossolo de cerrado em plantio de espécies florestais [Physical, chemical and biological attributes of a cerrado Latosol in planting forest species]. *Pesquisa Agropecuária Brasileira [Brazilian Agricultural Research]* 44: 613–620.

Silva, T. P., Nunes, M. C. M., Lima, C. L. R., Paiva, R.B., Barbosa, L. S., Fischer, L. O., 2018. Qualidade estrutural de um Argissolo Vermelho-Amarelo sob diferentes usos no Bioma Pampa no Sul do Brasil [Structural quality of a Red-Yellow Argisol under different uses in the Pampa Biome in Southern Brazil]. *Revista Brasileira de Agropecuária Sustentável [Brazilian Journal of Sustainable Agriculture]* 8: 115-123.

Six, J., Bossuyt, H., Degryze, S., Denef, K., 2004. A history of research on the link between (micro)aggregates, soil biota, and soil organic matter dynamics. *Soil and Tillage Research* 79: 7–31.

Six, J., Elliott, E. T., Paustian, K., 2000. Soil macroaggregate turnover and microaggregate formation: A mechanism for C sequestration under no-tillage agriculture. *Soil Biology and Biochemistry* 32: 2099–2103.

Six, J., Elliott, E. T., Paustian, K., Doran, J. W., 1998. Aggregation and Soil Organic Matter Accumulation in Cultivated and Native Grassland Soils. *Soil Science Society of American Journal* 62: 1367–1377.

Society for Ecologial Restoration Ser. *Princípios da SER International sobre a restauração ecológica [SER's International Principles on Ecological Restoration]*, 2004.

Stotzky, G. Influence of Soil Mineral Colloids on Metabolic Processes, Growth, Adhesion, and Ecology of Microbes and Viruses. In: Huang, P. M., Schnitzer, M. (Eds.). *Interactions of Soil Minerals with Natural Organics and Microbes* Soil Science Society of America, 1986.

Tisdall, J. M., Oades, J. M., 1982. Organic matter and water-stable aggregates in soils. *Journal of Soil Science* 33: 141–163.

Tótola, M. R., Chaer, G. M. Microrganismos e processos microbiológicos como indicadores da qualidade dos solos [Microorganisms and microbiological processes as indicators of soil quality]. In: Venegas, V. H. A. (Ed.). *Tópicos em Ciência do solo [Topics in Soil Science]*. Viçosa: Sociedade Brasileira de Ciência do Solo, 2002.

Totsche, K. U., Amelung, W., Gerzabek, M. H., Guggenberger, G., Klumpp, E., Knief, C., Lehndorff, E., Mikutta, R., Peth, S., Prechtel, A., Ray, N., Kögel-Knabner, I., 2018. Microaggregates in soils. *Journal of. Plant Nutrition and Soil Science* 181: 104–136.

Tucker Serniak, L. The Effects of Earthworms on Carbon Dynamics in Forest Soils. *Reference Module in Earth Systems and Environmental Sciences*. Elsevier Inc, 2017.

Vezzani, F. M., Mielniczuk, J., 2009. An overview of soil quality. *Revista Brasileira de Ciência do Solo [Brazilian Journal of Soil Science]* 33: 743–755.

Wagai, R., Kajiura, M., Asano, M., 2020. Iron and aluminum association with microbially processed organic matter via meso-density aggregate formation across soils: Organo-metallic glue hypothesis. *Soil 6*(2):597–627.

Wright, J. P., Jones, C. G., 2006. The concept of organisms as ecosystem engineers ten years on: progress, limitations, and challenges. *Bioscience* 56: 203–209.

Yoder, R. E. 1936. A Direct Method of Aggregate Analysis of Soils and a Study of the Physical Nature of Erosion Losses. *Agronomy Journal* 28: 337-351.

Zhang, W., Hendrix, P. F., Dame, L. E., Burke, R. A., Wu, J., Neher, D. A., Li, J., Shao, Y., Fu, S., 2013. Earthworms facilitate carbon sequestration through unequal amplification of carbon stabilization compared with mineralization. *Nature Communications* 4: 2576.

Chapter 5

The Drivers of Soil Organic Matter Accumulation

Clever Briedis[1,*]
João Carlos de Moraes Sá[2]
and Ademir de Oliveira Ferreira[3]
[1]Department of Agronomy, Federal University of Viçosa, Viçosa, Minas Gerais, Brazil
[2]Department of Soil Science and Agricultural Engineering,
State University of Ponta Grossa, Ponta Grossa, Paraná, Brazil
[3]Department of Agronomy, Federal Rural University of Pernambuco, Recife, Pernambuco, Brazil

Abstract

Soil organic matter (SOM) is a pivotal component for sustainable agriculture in any environment. SOM plays a crucial role in soil quality by providing positive effects on soil chemical, physical and biological properties besides increasing crop productivity. Furthermore, SOM is a component of the global carbon cycle, and changes in its content can mitigate or intensify the emission of greenhouse gases. High SOM content can ensure food security for a growing world population while reducing the adverse effects of climate change. Therefore, understanding SOM dynamics is decisive while creating strategies that enhance its accumulation and permanence in the soil. In general, these strategies should be based on increasing inputs and decreasing losses of SOM. Several natural (climate and soil) and anthropogenic (management practices) factors, in a dynamic interaction, drive the accumulation of SOM. The main climate factors are temperature and precipitation, which

* Corresponding Author's Email: clever.briedis@ufv.br.

In: Agricultural Soils
Editors: R. Nogueira de Sousa, T. Nascimento Pessoa and K. Ferreira Mendes
ISBN: 979-8-89113-000-5

can be associated with net primary productivity and SOM decomposition rates. Soil factors are mainly texture, mineralogy, and aggregation, having the most effect on SOM stabilization processes. Finally, several management practices can be used as a strategy for SOM accumulation, including smart land use, organic amendments, biochar, mineral fertilization, liming, cover crops, and no-till. Worthwhile highlight that management practices can be changed and improved, contrary to climate and soil, whose characteristics are immutable in the short term. Thus, this chapter emphasized and discussed management practices with more emphasis.

Keywords: SOM dynamics, SOM stabilization, management practices, soil quality

Introduction

Soil organic matter (SOM) is the organic component of soil that consists of plant, microbial, and animal tissues in various stages of decomposition. Despite comprising just 1–5% of the soil mass, SOM is considered the basis for achieving the sustainability of any agroecosystem, as it directly affects the quality, health, and functionality of soil (Lal, 2016). The beneficial effects of SOM can be observed on soil physical (improvement of soil structure, increasing water retention, decreasing bulk density), chemical (increasing cation exchange capacity, buffering changes in soil pH, supplying nutrients to plants), and biological properties (being the substrate for microorganisms), with a consequent increase in the productive capacity of the soil (Lal, 2020; Weil and Magdoff, 2004). In addition, SOM is the largest terrestrial reservoir of soil organic carbon (SOC), which can contribute enormously to the reduction or increase of atmospheric CO_2, the most relevant greenhouse gas (Lal, 2004). Therefore, agronomists and environmentalists should be concerned about increasing SOM levels in agricultural soils since high SOM content may guarantee food security and delay the climate change effects.

SOM dynamics are very complex, being governed by several factors such as climate (e.g., temperature and precipitation) (Burke et al., 1989; Chen et al., 2020), soil (e.g., texture, mineralogy, and aggregation) (Dungait et al., 2012; Schmidt et al., 2011), and management practices (e.g., tillage system, use of amendments and fertilizers, land use systems, and cover crops) (Briedis et al., 2018; De Oliveira Ferreira et al., 2021; Romaniw et al., 2021). The interaction of these different factors controls the balance of inputs (above and below ground) and outputs, driving the SOM accumulation or loss over time (De Moraes Sá et al., 2022). Understanding this dynamic is critical to designing SOM accumulation strategies more efficiently.

This chapter concentrates on addressing the drives of SOM accumulation in agricultural soils. At first, the mechanisms of SOM stabilization are reviewed, and the climatic and edaphic factors and management practices that drive the SOM accumulation are presented and discussed in the sequence. Finally, as SOM is approximately 58% C, measured SOC serves as a proxy for SOM, so SOC will often be used in this chapter as a synonym for SOM.

Soil Organic Matter Dynamics

SOM has complex dynamics in natural and agricultural environments, but in a simplified way, it is the net balance between organic matter inputs and losses (Figure 1). Thus, land management must be driven to increase inputs and decrease outputs. The inputs are mainly represented by senescent plant residues (shoots and roots) produced on-site and by root exudates or external sources of organic material (e.g., manure, organic wastes, organic amendments). On the other hand, the outputs occur mainly by microbial decomposition but may also be by erosion, leaching, and burning (Chenu et al., 2019).

Both inputs and outputs of organic matter are controlled by natural factors such as climate and soil but can also be altered by management practices (Figure 1). In agricultural environments, the increase in inputs is related to a higher net production of an area, which can be achieved by an effective crop rotation, including cover crop species with a high biomass production capacity (especially roots) (Gonçalves et al., 2019; Perkons et al., 2014). In addition, other practices must be implemented to achieve high biomass production, such as using cash crops with high productive potential, adequate fertilization, use of organic fertilization, control of soil acidity, irrigation, etc. In tropical environments, the minimum range of biomass to maintain the SOM at a dynamic equilibrium level is estimated to be 12–15 Mg ha^{-1} yr^{-1} (Sá et al., 2015), while at the subtropics this range is found at approximately 8–10 Mg ha^{-1} yr^{-1} (De Oliveira Ferreira et al., 2012), thus being an uppermost challenge for farmers.

The decrease in outputs depends on conservationist agricultural practices that eliminate burning, limit soil erosion, and control the rate of microbial decomposition. The regulation of microbial decomposition rates, which allows SOM persistence, is a process governed by the biotic and abiotic environment, which regulates the access of microbial decomposer to SOM, and not specifically by the molecular composition of SOM (Schmidt et al., 2011). Furthermore, microbial activity regulates not only SOM decomposition but also SOM formation and stabilization. Recently, microbial necromass has been postulated to contribute over 50% of total SOM in croplands (Angst et al., 2021; Liang et al., 2019). Thus, the dynamics of increase and permanence of SOM over time depend both on inputs via residues and on environmental characteristics that promote its stabilization, which is mainly governed by microorganisms (Figure 1).

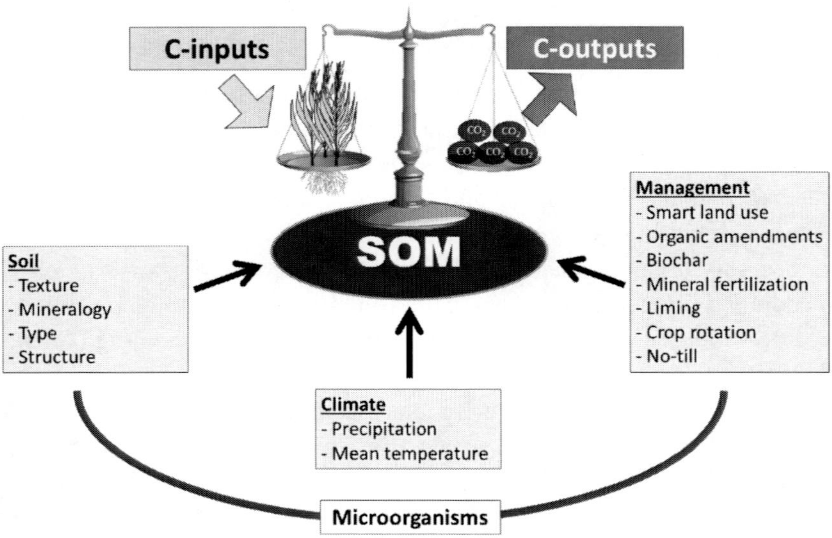

Figure 1. Diagram summarizing the dynamics and main drivers of soil organic matter (SOM) accumulation in the agricultural environment. SOM dynamics involves the net balance between C-inputs and C-outputs. Several soil, climate, and management factors drive these two parameters and thus SOM accretion in soil. Furthermore, microorganisms may regulate both the decomposition and formation of SOM.

Mechanisms of SOM Stabilization

Considering that, to a large extent, the dynamics of SOM are governed by its decomposition rate, understanding the SOM stabilization against microbial degradation is essential to guide agricultural management to prolong its permanence and accumulation over time. In mineral soils, SOM stabilization is governed by its chemical recalcitrance (biochemical stabilization), its sorption with minerals (chemical stabilization), or its physical protection by aggregates (physical stabilization). The relative importance of each of these mechanisms in protecting SOM may vary with climate, soil texture and mineralogy, and agricultural management (von Lützow et al., 2006). In addition, it is worth noting that microorganisms are currently considered to play a fundamental role in SOM stabilization (Angst et al., 2021; Cotrufo et al., 2013; Liang et al., 2017; Liang et al., 2019), mainly interacting with the chemical stabilization of necromass and secondary metabolites with soil minerals (Angst et al., 2021; Liang et al., 2019), in addition to being encapsulated within soil microstructures (Totsche et al., 2018).

For a long time, biochemical stabilization was considered the principal mechanism controlling SOM decomposition. Through this process, the SOM degradability is reduced because of the chemical and structural stability of its biomolecule (Krull et al., 2003). SOM protection by this mechanism, also known as selective preservation, initially depends on the indigenous molecular characteristics of plant litter and the recalcitrance of humic polymers formed by condensation processes. In this sense, labile substances (e.g., polysaccharides and proteins) from plant residues are preferentially degraded during the decomposition process, and polymers that contain aromatic rings, such as in lignin, are selectively preserved (Krull et al., 2003). However, isotopic studies have demonstrated a rapid turnover even of organic molecules with the most recalcitrant structure (Rasse et al., 2005), and the persistence of such molecules depends very much on the presence and access of suitable decomposer organisms (Schmidt et al., 2011). In the early stages of decomposition, selective preservation is still considered of relative importance for SOM protection (Marschner et al., 2008).

Currently, chemical stabilization governed by the adsorption and chemical binding of organic molecules with soil mineral surfaces is considered the principal mechanism for long periods of SOM preservation (Kögel-Knabner et al., 2008; Wiesmeier et al., 2019). The chemical interactions of SOM on mineral surfaces can occur through various mechanisms, such as ligand exchange, cationic bridges, and weak interactions (van der Waals forces and H-bonding) (von Lützow et al., 2006). Once the organic molecule is bound with soil minerals, the accessibility of the microbial enzymes responsible for SOM breakdown is decreased (Kleber et al., 2015). Thus, long SOM turnover times have been observed in silt and clay fractions (i.e., MAOM – mineral-associated organic matter) (Marschner et al., 2008) and linear relationships between SOM content and the content of these minerals in the soil (Georgiou et al., 2022; Hassink, 1997; Six et al., 2002). Currently, it is considered that the MAOM fraction is formed mainly by microbial material attached to the mineral surface (Cotrufo and Lavallee, 2022).

Not only soil texture interfere with SOM chemical protection, but also the type of clay mineral (Feng et al., 2013). In general, minerals with high specific surface area (SSA) have abundant reactive surfaces which SOM can be adsorbed, thus having a high SOM protection capacity (Wiseman and Püttmann, 2005). Variations in soil mineralogy, however, can alter the binding mechanism of SOM with minerals. In 2:1 layer type phyllosilicates minerals (e.g., vermiculite, illite, and smectite) with permanent negative charges, the main form of bonding is through cationic bridges, with Ca^{2+} being the principal polyvalent cation in this bridge (Loganathan et al., 2020). On the other hand, iron (Fe) and aluminum (Al) oxides provide large reactive surface areas and may have a higher impact on SOM stabilization than clay minerals (Mikutta et al., 2006; Souza et al., 2017). Indeed, the SOM association with these metallic sesquioxides occurs preferentially by exchanging ligands, which is a stronger form of bonding when compared to cationic bridges (Kleber et al., 2015). Thus, highly weathered soils from the tropics, rich in minerals such as goethite (Fe oxide) and gibbsite (Al oxide), naturally promote high SOM protection against microbial degradation (De Oliveira Ferreira et al., 2018a; Zinn et al., 2007).

In addition to chemical stabilization, physical stabilization, which is conferred by soil aggregates, is another significant way of delaying SOM turnover. The occlusion and encapsulation of SOM within aggregates decreases its mineralization since it reduces access to the microorganisms and their enzymes and restricts aerobic decomposition due to the diminished diffusion of oxygen (Kravchenko and Guber, 2017). However, there is a difference between the protective capacity of SOM within microaggregates (<250 μm) compared to macroaggregates (>250 μm). While SOM in microaggregates is estimated to have a turnover time of centuries, it has a turnover time of a few years when inside macroaggregates (Puget et al., 2000). Thus, the effectiveness of the physical stabilization of SOM depends on the formation and stability of microaggregates. For this reason, soils with higher clay and Al- and Fe-oxide contents provide more highly stable microaggregates, making them very effective in protecting SOM (Totsche et al., 2018).

Despite the lower protection capacity, macroaggregates play a crucial role in SOM accumulation over a long period. Physical protection promoted by macroaggregates is the first step for labile SOM to be incorporated within microaggregates for long-term stabilization (Tivet et al., 2013b; Totsche et al., 2018). In addition, the physical protection of labile SOM by the macroaggregate allows a longer time for interactions between mineral particles and microbial residues to be established, resulting in the formation of a very stable organo-mineral complex (Plaza et al., 2013). However, as macroaggregates are highly affected by agricultural management, it is essential to constantly maintain conditions for increasing their formation and decreasing their turnover (De Oliveira Ferreira et al., 2018b; Six et al., 2000).

Climate as Driver of SOM Accumulation

Climate is considered a dominant factor that explains the SOM distribution at regional to global scales, especially on the soil surface (i.e., 0-20 cm) (Luo et al., 2021). The climate variables of mean temperature and precipitation could regulate the biome, the type of vegetation, the quantity and quality of the residues, the microbial activity, and the rates of SOM decomposition (Burke et al., 1989). Thus, they can drive both the SOM inputs and outputs. However, at a local scale, management variables (e.g., irrigation, soil tillage, fertilization) may mask the effects of climate and play a big role in SOM accumulation.

The relationship between temperature and precipitation with SOM stocks is not always straightforward and may have several interactions. In general low rainfall can limit the amount of above- and below-ground biomass production, while high temperatures can increase microbial activity and thus accelerate SOM decomposition (Epstein et al., 2002; Plaza et al., 2018). On the other hand, in regions without hydric restriction, the temperature can be the principal driver, where low temperatures slow down microbial activity and thereby preserve SOM (Schimel et al., 1994). Furthermore, the high precipitation intensity is an important regional factor to be evaluated since it can aggravate SOM losses by erosion (Nearing et al., 2005). Taking all this into account, on a macro scale, soils from arid regions generally have lower SOM contents than soils from humid and cool ones.

On a long timescale, climate can still interact with SOM stabilization processes. High precipitation and temperature in tropical climates accelerate the weathering process of the parent material, increasing the amounts of Fe and Al oxides, which in turn enhances the chemical stabilization (i.e., association with minerals) of SOM (von Lützow et al., 2006). While the direct relationship of climate (temperature and precipitation) with SOM is greater at the soil surface (Luo et al., 2021), the indirect relationship by determining soil texture and mineralogy tends to increase with soil depth due to the greater importance of SOM stabilization through organo- mineral associations in deeper layers (Poirier et al., 2020).

We must not forget that the planet is experiencing climate change, with rising global temperatures and more frequent severe weather events (e.g., droughts and high-intensity rains). This new climate status can alter the SOM persistence and accumulation, especially on the soil surface. These relationships between climate change and SOM are complex and not well understood. In general, SOM losses can be accelerated as temperature increases since the rate of microbial decomposition also accelerates. However, higher plant production by fertilization with increased atmospheric CO_2 can counterbalance the possible SOM losses by increasing temperature. Again, the balance between increased C inputs due to CO2 fertilization and C outputs due to increased SOM decomposition will determine SOM dynamics.

Soil as Driver of SOM Accumulation

In contrast to climate, which may serve as an indicator for SOM content at larger scales (regional to global scales), soil characteristics influence local to subregional scales. The main soil characteristics influencing SOM accumulation are texture, mineralogy, type, and structure. Considering the texture, the clay content (<2 μm) is the principal determinant since it provides the surface for SOM adsorption. As previously mentioned, microorganisms and their enzymes cannot access SOM bound to mineral surfaces, which is the major mechanism for long-term SOM preservation (Kleber et al., 2015). In addition, soil texture highly influences the pore network and microstructure and the formation of microaggregates, which increases the SOM physical protection (Totsche et al., 2018). Thus, SOM losses due to inappropriate management practices are usually higher in coarse than in finer textured soils (Dieckow et al., 2009). Despite its great importance, there is a limit of SOM content that clay minerals can retain, leading to the process known as soil C saturation (Briedis et al., 2016; Hassink, 1997; Six et al., 2002). In a study carried out with soils from tropical and subtropical regions, Briedis et al. (2018) showed an asymptotic relationship between C in bulk soil and C associated with silt+clay minerals, indicating that the soil mineral fraction tends to become C-saturated. However, the authors observed that the current

SOC levels were still far from the estimated saturation value, with a potential for SOM accumulation with continued conservation practices.

Soil mineralogy is another determining factor in SOM accumulation since it is directly related to the reactive surface of minerals. Each type of mineral (1:1-layer phyllosilicates, 2:1-layer phyllosilicates, and Fe/Al oxides) has different characteristics of reactive surfaces sites, exerting differences in the sorption capacity of SOM (Han et al., 2016a). For example, smectites and vermiculites (2:1-layer phyllosilicates) have higher cation exchange capacity and specific surface than kaolinites (1:1-layer phyllosilicate), therefore having higher sorption capacity (Singh et al., 2018). Thus, soils from temperate climate regions higher in 2:1 minerals would theoretically have a greater protective capacity and higher SOM content. However, this dynamic changes when Fe- and Al- oxides are present, like in highly weathered soils of tropical and subtropical regions. Firstly, the adsorption of SOM with these minerals is energetically very strong since interactions occur mainly via ligand exchange (Mikutta et al., 2006). Furthermore, Fe- and Al- oxides can form stable microaggregates by the cementation of other minerals and cause the flocculation of clay, resulting in the occlusion of SOM at the microstructure level (Totsche et al., 2018). Thus, different soil mineralogies, governed by differences in the ecosystem, largely determine the SOM accumulation potential of a given soil. For example, the high Fe and Al oxides content in the highly weathered soils (Oxisols) associated with the higher elevation and low temperatures promoted a prone condition to the C protection (De Oliveira Ferreira et al., 2018a). Other studies conducted in subtropical soils also indicate that the C accumulation in subsoil layers is primarily mediated by Fe and Al oxides (Reis et al., 2014; Rumpel and Kögel-Knabner, 2011; Zinn et al., 2007).

Soil aggregation has also been highlighted as a pivotal agent in SOM accumulation (Briedis et al., 2012a; De Oliveira Ferreira et al., 2018b; Tivet et al., 2013b). Furthermore, aggregation becomes very important in agricultural soils because they may be altered by management practices, unlike soil texture and mineralogy. Soil aggregates consist of the union of primary soil particles in small structures that are formed through the action of cementing agents such as gelatinous chains of polysaccharides (transient agents), plant roots, fungal hyphae (temporary agents) and also by the action SOM, Fe, Al, and Ca (persistent agents) (Tisdall and Oades, 1982). Aggregates protect SOM against decomposition by microbial biomass since they separate the organic substrate from microorganisms as well as control the diffusion of oxygen into aggregates (Kravchenko and Guber, 2017; Totsche et al., 2018). Although long-term SOM stabilization is more related to microaggregates, the low turnover of macroaggregates is pivotal for the stabilization of microaggregates within it (Six et al., 2000). The way to increase macroaggregation go through minimal soil disturbance, continuous deposition of plant residues, and permanent soil cover, which creates a constant flow of labile compounds of biological origin (cementing agents) capable of efficiently stabilizing the macrostructure (De Oliveira Ferreira et al., 2018b; Tivet et al., 2013a). In addition, the higher the stability of macro aggregates, the better the structural quality, which improves root growth, water infiltration, and aeration, resulting in increased plant production and thus indirectly affecting the SOM accumulation. Finally, more structural quality protects against SOM losses by erosion processes.

Management Practices as Drivers of SOM Accumulation

On a local and farm scale, management practices highly affect the SOM dynamics, especially on the soil surface. Management strategies should always consider the net effects of C inputs and outputs, prioritizing the increase in below- and above-ground biomass production and the decrease in microbial decomposition and erosion. These practices include smart land use, organic amendments, biochar, mineral fertilization, liming, crop rotation, and no-till, which will be discussed below.

Smart Land Use

Land use change is one of the main factors of SOM alterations. Globally, the conversion of native vegetation to agricultural lands has been responsible for about 40% of native SOC losses (Wei et al., 2014). These SOC losses happen for the decrease in C-inputs by vegetation and the increase in decomposition rates (Briedis et al., 2021; De Moraes Sá et al., 2022).

Clearing native vegetation and replacing it with crops generally promotes fewer residues returned to the system because to export through harvesting. In addition, modern agricultural cultures have less rooting, which reduces the contribution of C-belowground. It is currently agreed that C from roots (exudates and dead roots) is the main source for SOM formation (Rasse et al., 2005).

The increase in decomposition rates, in turn, occurs mainly due to the use of systems with intense soil mobilization (Sá et al., 2014). Soil preparation for planting through plowing and harrowing promotes the breakdown of aggregates that were previously protecting SOM. The incorporation of fresh organic compounds, combined with the exposure of protected SOM and increased aeration, accelerates microbial activity and causes intense SOM decomposition (Briedis et al., 2018; Sá et al., 2015). In addition, the breakdown of the structure and soil spraying by tillage makes it easier for SOM to be lost by erosion.

Land conversion from native land to poorly managed agricultural areas has a high potential for SOM losses (Friedlingstein et al., 2020). However, smart land use can contribute as a meaningful strategy to recover part of historical SOM losses. One of these land use strategies is agroforestry systems. These systems involve the integration of different combinations of trees with crops and/or animals, aiming at the ecological and economic interactions of the distinct components. In this integration of forest and agriculture, several benefits are provided, as well as noticeable efficiency in water, nutrients, and light use (Lorenz and Lal, 2014b). This system also promotes greater income diversification for farmers through crops, fodder, and firewood/timber, in addition to other environmental benefits such as improved soil fertility, increased microbial biodiversity, erosion control, increased species richness, increased nutrient cycling, and increased water quality and availability (Jose, 2009; Muchane et al., 2020).

From the view of increasing SOM, agroforestry systems are also eminent, both on the surface and underground. One of the main factors for the prominent potential of SOM accumulation in agroforestry systems is the high production of aboveground biomass (stem, branch, and foliage) and deep root systems of trees in that system (Shi et al., 2018). Particular attention should be given to the more and deeper roots in this agroforestry compared to crops in monoculture. Root-derived C contributed 1.5 to 3 times more to SOM stabilization than shoot-derived C (Lorenz and Lal, 2014b). Thus, agroforestry systems have a great capacity to stabilize C in the subsoil by chemical interactions with soil minerals, slowing down the turnover of the new incorporated SOM.

Aside from increased C inputs, especially roots and exudates, agroforestry systems also have a high potential for SOM accumulation due to greater erosion control (Zhu et al., 2020). This occurs due to the non-disturbance of the soil and the permanent maintenance of soil cover in this system. Another mechanism for higher SOM accumulation in agroforestry systems is the ability of plants to recover nutrients (nitrogen, phosphorus, calcium, magnesium, and potassium) from below the crop rooting zone (Jose, 2009). In addition to C, these nutrients are essential for the formation and stabilization of SOM (Kirkby et al., 2014).

Although agroforestry systems have a high potential for SOM accumulation, this can vary depending on the type of agroforestry system, tree species, climatic conditions, soil conditions, and management practices. In a recent meta-analysis, De Stefano and Jacobson (2018) showed that using agroforestry systems can increase SOM by 26% in the 0-30 cm layer. However, the results varied according to the type of the previous crop, having a better potential for gain in the transition from agriculture to agroforestry (40% increase) than pasture/grassland to agroforestry (9% increase).

Another smart land use with a high potential for SOM accumulation is the integrated crop-livestock system. This system integrates pasture, livestock, and crop production activities in the same area, utilizing on-site animal services as a resource for crop production and cropland to support livestock production (Brewer and Gaudin, 2020). This system is still planned to achieve synergy and emergent properties resulting from the spatial and temporal interactions among the components of soil, plant, animal, and atmosphere (de Moraes et al., 2014). The abundant benefits of integrated systems include improving soil structure, water infiltration, nutrient cycling, SOC sequestration, and soil biological diversity (Salton et al., 2014).

Compared to crop fields or degraded pastures, integrated systems have demonstrated substantial potential for SOM accumulation. In a study conducted in southeastern Brazil, Bieluczyk et al. (2020) identified that the change from an extensive low-grazing intensity pasture to an integrated crop-livestock system led to an annual accumulation of 0.28 Mg C ha^{-1} from 2010 to 2016. Higher SOM accumulation rates have been observed when integrated systems are compared with agricultural areas. Studying land use changes in the Amazonia and Cerrado biomes, (Carvalho et al., 2010) detected SOC accumulation ranging from 0.82 to 2.85 Mg ha^{-1} $year^{-1}$ when agricultural areas were converted to integrated systems. According to the authors, the highest accumulation rates were associated with the high soil fertility, local climate conditions, and the presence of forage for a longer time in the area.

The high potential of integrated systems to accumulate SOM is mainly due to the increase in plant production in grazed areas compared to that in non-grazed ones, this effect being subject to grazing intensity. Light and moderate grazing stimulate the production of tillers and roots by shoot renewal, besides removing senescent tissues, increasing light availability for actively photosynthesizing vegetation (Assmann et al., 2014).

Organic Amendments

Different forms of organic amendments have been used for thousands of years to improve the productive potential of soils. The two principal organic amendments used in agricultural lands are animal manure and crop residues. Compost, municipal biosolids, and waste from manufacturing processes still have some importance (Scotti et al., 2015). These products can be considered soil conditioners, affecting plant growth indirectly by improving the physical and biological properties of the soil, but also as a source of nutrients. The application of organic amendments can have a direct effect on the increase of SOM, which is generally proportional to the amount of organic C applied, and an indirect effect by increasing the addition of C from crop residues since it improves the productive capacity of the soil (Romaniw et al., 2021). In a recent meta-analysis, Bolinder et al. (2020) demonstrated a potential for SOC accumulation by using organic amendments ranging from 0.2 to 1.3 Mg C ha^{-1} $year^{-1}$. According to the authors, the SOC gain potential varied with the application rate and the quality of the added material. Furthermore, the potential for SOM increase with continuous application of organic amendments tends to decrease over time, reaching a new equilibrium after a few decades, probably due to the soil C saturation phenomena (Six et al., 2002). However, additional SOC can still be captured to the underlying soil layer through the translocation of soluble C after saturation of the topsoil layer (Nicoloso et al., 2018).

Organic amendments application can still lead, in some circumstances, to an increase in the rate of SOM decomposition and a decrease in its amount in the soil. This is related to a possible increase in microbial activity (priming effect) after using materials with relatively high amounts of organic nitrogen (Quilty and Cattle, 2011). Therefore, care must be taken to understand the composition of the organic amendment that is added to the soil.

Biochar

Biochar is the product of pyrolysis (partial combustion under low oxygen conditions) of organic biomass and has gained massive attention lately (Weber and Quicker, 2018). The increased research interest in using pyrolyzed products in agricultural soils was triggered by the discovery of "Terra Preta" (black earth), an anthropogenically modified Brazilian Oxisols (originally of low fertility) capable of providing high yields relative to the adjacent soil (Lehmann et al., 2003). The greater productive capacity with the addition of biochar is due to increases in total carbon, nutrient availability, soil moisture holding capacity, and promoted microbial activity (Zhou et al., 2021).

Several materials can serve as feedstock for biochar production, including woody residues, food and green wastes, biosolids, and animal manures. The result of pyrolysis is the stabilization of a significant portion of C, which increases the recalcitrance of the produced biochar. Around 95% of biochar consists of highly stable aromatic C compounds (Wang et al., 2016). This causes a reduction in its decomposition rate and therefore has an expectation of permanence in the soil on a centennial scale, much higher when compared to unburnt organic material from which it originates (Lehmann et al., 2015). However, some factors change the recalcitrance of the final product, highlighting the initial biomass feedstock and the pyrolysis temperature. Generally, biochar made of woody material has a longer residence time than biochar from manure or crop residues because of the higher amount of lignin from the first. Likewise, high pyrolysis temperatures (e.g., 700−900 °C) generate biochars with a longer half-life than when produced at low temperatures (e.g., 200−400 °C), mainly by increasing the proportion of aromatic C and the higher degree of aromatic condensation (Singh et al., 2012; Wang et al., 2016).

In a recent global meta-analysis evaluating field experiments between 1 and 10 years with biochar application amounts between 1 and 100 Mg ha^{-1}, Gross et al. (2021) observed a significant increase of 13.0 Mg ha^{-1} of SOC with the application of biochar, which represented a percentage gain of 29% compared to the treatments that did not receive the product. The SOM increase through the biochar application goes beyond the direct introduction of a stable source of C as soon as this pyrolyzed product can also delay the SOM mineralization, a process known as the negative priming effect. This is due to a change in the behavior of soil microorganisms, which have preferential use of easily available C from incomplete pyrolysis in biochar and decreasing the decomposition of native SOM. (Wang et al., 2016). Furthermore, the increase in SOM after biochar application may occur due to the indirect effects of increased C input from higher plant production in soils of improved quality (Lorenz and Lal, 2014a). Although most studies point to a positive effect on SOM accumulation by using biochar, losses can also occur, mainly due to the positive priming effect. This effect depends on factors such as feedstock type, pyrolysis temperature, soil texture, and organic matter level of the soil (Majumder et al., 2019; Wang et al., 2016).

Mineral Fertilization

To achieve high production of food, feed, fiber, and bioenergy crop products, modern agriculture needs to use mineral fertilization (commonly containing nitrogen, phosphorus, and potassium) to replenish the nutrients exported by the harvest, thus maintaining soil fertility. Crops can double their yields when properly fertilized (Lu, 2020). However, to develop sustainable agriculture, it is essential to use these mineral fertilizers in a balanced way, always considering the ability of the soil to supply nutrients and the plant's nutritional needs. The fertilizer application from the right source, at the right dose, at the right application timing, and in the right place may enhance crop yield and nutrient use efficiency and decrease the harmful potential of synthetic fertilizer use (Johnston and Bruulsema, 2014).

While increasing the primary production and thus the amount of above- and belowground litter added to soil, balanced mineral fertilization can indirectly increase the amount of SOM over time. In a global meta-analysis of 298 different sites, Han et al. (2016b) demonstrated that the application of NPK together with the return of residues had a significant positive effect on the SOC sequestration rate (0.37 g kg^{-1} yr^{-1}) on the cropland topsoil (i.e., 0–20 cm) when compared to the unfertilized control. However, when mineral fertilization was unbalanced, SOC sequestration was only 0.07 g kg^{-1} yr^{-1}. The results also showed a strong correlation between the amount of residue added, higher in fertilized areas, with the SOC stock. Another study using seven different long-term experiments also demonstrated a positive effect of NPK fertilization on SOC sequestration (0.12-0.22 Mg ha^{-1} yr^{-1}) compared with the unfertilized control (Poeplau et al., 2018). However, in the study, belowground and aboveground inputs could not fully explain the differences between treatments with and without fertilizers, demonstrating the existence of other mechanisms of SOM accumulation in fertilized areas.

It is known that to form SOM, in addition to C, microorganisms need nutrients such as N, P, and S. In soils with a limitation of these nutrients, microorganisms, to decompose the added residue, can mine old SOM to meet their nutrient requirements, thus promoting a priming effect (Chen et al., 2014; Coonan et al., 2020). Thus, nutrient supplementation via mineral fertilization can enhance the accumulation of SOM. In this sense, Kirkby et al. (2013) reported an increase of two to eight folds in the SOC accumulation efficiency by N, P, and S addition compared to the control. Likewise, Briedis et al. (2016) demonstrated greater efficiency of SOC accumulation in topsoil (0-20 cm), which was higher in P, Ca2+, and Mg2+ availability, highlighting that nutrient scarcity in the subsoil can limit SOC accretion, even deeper soil having substantial potential for SOC sequestration due to its low level of C saturation.

In addition to decreasing the priming effect by lower microbial nutrient mining, the addition of nutrients can lead to the formation of better quality residues and thereby increase microbial anabolism and the efficiency of C utilization by these decomposer microorganisms (Poeplau et al., 2019). This is because the biosynthesis of decomposer cells requires a relation of C and nutrients in relatively stable ratios (Manzoni et al., 2012). When NPK is added to the soil, the residues C:N and C:P ratios get closer, approaching the ideal microbial synthesis ratio and thus increasing the stabilization of C in microbial necromass (Poeplau et al., 2018). The stabilization of necromass in the soil mineral fraction is currently considered the main form of long-term SOM accumulation and stabilization (Miltner et al., 2012).

Lime and Gypsum

Soils in many agricultural areas worldwide are acidic and have high Al^{3+} activity. Thus, agriculture in these places depends on lime and gypsum to achieve high crop yields. In addition to correcting problems associated with acidity and raising and improving soil fertility, these practices add the macronutrients Ca^{2+}, Mg^{2+}, and SO_4^{2-} ions. Ca^{2+} still enhances the root system when moved to the deeper soil layers (Caires et al., 2006). However, the outcome in the long-term of these practices on SOM dynamics may be controversial. Rising pH close to neutrality by liming increases microbial activity and allows short-term SOM decomposition. In addition, liming can promote the dispersion of clay colloids in acidic soils (Roth and Pavan, 1991) and accelerate the degradation of labile SOM (a binding agent for aggregation), leading to soil disaggregation and, thus decreasing physical protection of SOM (Tisdall and Oades, 1982).

Despite a possible initial loss, progressive SOM increments have been observed in long-term experiments. For example, Briedis et al. (2012a) demonstrated that, compared to no lime treatment, lime applied to reach the base saturation close to 70% promoted a SOC accumulation of 7.6 Mg ha^{-1} at

0-20 cm soil layer after 15 years, which represented an annual rate of 0.51 Mg ha^{-1}. One of the mechanisms to explain this increase in SOM was the beneficial effect of liming on biomass production (cash and cover crops), mainly roots. Caires et al. (2006) observed that the amelioration of soil fertility in subsoil layers has been primarily observed in long-term no-till soils. According to De Oliveira Ferreira et al. (2018a), to promote the increase of soil fertility in Oxisol's deeper layers, it is necessary to fulfill certain requirements. Among them: a) the buildup of a surface layer rich in basic cations, which will work as an eluviation layer. High amounts and frequency of Ca-based soil amendments are necessary to saturate this layer with Ca. b) the precipitation level needs to be high enough to provide water for cation transport along the soil profile. c) The K content on the surface layer needs to be elevated to avoid an imbalance of bases. d) The presence of continuous pores associated with the high biological activity is essential to provide water infiltration. e) It is necessary to use a crop rotation system with high biomass-C input and plants with a deep root system to form a soil layer rich in SOC (Nora and Amado, 2013). These actions associated with a minimum soil disturbance allow cation leaching, neutralization of toxic Al in subsurface layers, deep root system development, and increased SOC. These processes are supported by the linear models adjusted for the surface (0-15 cm) and subsurface (60-100 cm) soil layers (De Oliveira Ferreira et al., 2018a). The models highlighted the importance of increasing soil pH and keeping nutrient availability to plants aligned with high plant biomass input for support C recovery in surface layers.

In addition to the higher C-residues input, increases in soil aggregation may also be involved in SOM accretion with liming. In this sense, Briedis et al. (2012b) observed an increase in the diameter of aggregates with liming, indicating an increment in aggregation and SOM physical protection. The higher amount of roots reported in the liming treatment was mentioned as one factor for the improved stabilization of aggregates. When alive, roots physically join soil particles and release polysaccharides that stabilize aggregates (Tisdall and Oades, 1982). When dead, they can still be the nucleus for the formation of new macroaggregates (Six et al., 1999). Incidentally, (Briedis et al., 2012a) found a higher concentration of polysaccharides and microbial C (proxied by the hot water extracted C) in the limed treatment. Furthermore, where lime and gypsum are added, calcium can form a cationic bridge between clay particles and organic matter, increasing soil aggregation, mainly at the microaggregate level. This role of calcium at the microscale level can be pointed out as one of the principal mechanisms for the long-term effect of liming on soil structural stability.

In addition, Ca^{2+} may act as a bridge between negative charges of SOM and clay, providing direct chemical protection of SOM against microbial degradation. Classically, the interaction between Ca^{2+} and SOM is considered to be through outer sphere reactions, which are considered weak and short-lived. However, recent studies have considered the possibility of this reaction being an inner sphere, mainly with carboxylic groups, which makes the Ca-SOM interaction much stronger and long-lasting (Rowley et al., 2018). The high dependence of Ca on C at the molecular level was demonstrated through the analysis of soil aggregates by energy-dispersive X-ray spectroscopy (Briedis et al., 2012b). In their study, Briedis and collaborators demonstrated a higher relationship between Ca and C in the limed treatment ($R^2 = 0.70$) compared to the control ($R^2 = 0.51$). Furthermore, they observed a stronger molecular correlation between C with Ca than with Al ($R^2 = 0.43$) and Fe ($R^2 = 0.01$), two elements abundant in highly weathered soils such as those in their study. This result demonstrates that under natural soil conditions in the tropics, Fe and Al are the main cations that act as a cationic bridge, but in agricultural soils that have received lime and gypsum, Ca becomes the most one.

In summary, lime and gypsum at adequate doses increase biomass production, increment soil aggregation, and strengthen Ca-SOM interaction, thus becoming an essential tool for SOM accumulation in acid agricultural soils.

Cover Crops

Cover crops consist of growing plants between or intercropped with cash crops to provide multiple benefits, such as preventing soil erosion, improving cycling nutrients, fixing atmospheric N (when leguminous species are used), suppressing weeds, controlling pests and diseases, increasing water infiltration, reducing drought stress, and improving soil structure (Adetunji et al., 2020). With the general improvement in soil quality, using cover crops also results in higher yields in cash crops grown in the sequence (Garbelini et al., 2022). Furthermore, in a crop rotation system, they can be used to increase the input of above- and belowground biomass to the area. This is because most cash crops (e.g., soybeans, cotton, beans, and wheat) return little C-residue to the soil, especially roots. After all, modern genotypes are developed to enhance photosynthate allocation to harvested organs vs. shoot or roots. Due to their large amounts of plant biomass, cover crops play a central role in the SOC accumulation in agricultural soils. In a meta-analysis conducted with studies from temperate regions, higher SOC accumulation by using cover crops was observed in 79% of the experiments, with an average relative increase of 12% compared to the control treatments. However, these differences rose to 30% under high cover crop biomass production (> 7 Mg ha^{-1} yr^{-1}), demonstrating that the greater the capacity to contribute C-residues by a cover crop, the greater it's capacity to SOM increase (McClelland et al., 2021). In a meta-analysis covering 139 plots at 37 sites spread globally, Poeplau and Don (2015) showed a SOC accumulation rate of 0.30 Mg ha^{-1} yr^{-1} using cover crops. In addition, the authors observed a linear relationship between the time of use of cover crops and changes in SOC stock. The authors also showed no evidence of C saturation even after 54 years of cultivation of cover crops.

Among the factors for SOM accumulation by using cover crops, the high contribution of belowground C (roots and rhizodeposits) can be considered one of the main ones. In fact, roots have between 1.5 and 3.7 times higher retention of carbon in the medium and long term than the aboveground biomass (Paul et al., 1999). This extensive contribution of roots in the SOM accumulation may be associated with some mechanisms. First, deep-rooting species can transfer root exudates (e.g., sugars, amino acids, and other organic acids) to the subsoil, where a high potential for SOM accumulation exists (Briedis et al., 2016; Briedis et al., 2018) and where these exudates are directly adsorbed to the mineral surface (Kaiser and Kalbitz, 2012). In addition, the roots and their exudates serve as a substrate that stimulates microbial growth, increasing the interaction and sorption of microbial necromass to the clay fraction (Angst et al., 2021; Cotrufo et al., 2015; Cotrufo et al., 2013). Finally, roots (fine roots, root hairs), exudates, and hyphae of root-associated fungi can be physically protected by direct entry into micrometer-scale pores and microaggregate (Rasse et al., 2005). Thus, management that enhances belowground C inputs should be prioritized when a system is thought to increase SOM content. An example of how to increase belowground inputs is the inclusion of deep-rooting perennials that develop roots in layers below 100 cm, such as Urochloa ruziziensis, used in production systems involving intercropping such as maize in Central Brazil. This strategy is used to keep the soil covered throughout the end of the rainy season, the dry period until the beginning of the new rainy season, and has contributed to the increase of SOM in no-till areas in Brazil (Sá et al., 2015).

In addition to the amount of above and belowground residues, crop diversity can also affect SOM because of the difference in the quality of organic material entering the soil. In a global meta-analysis, Jian et al. (2020) observed that when cover crops were included in rotations, SOC increased by an average of 15.5% across all experiments (from 131 publications) with an overall mean C sequestration rate of 0.56 Mg ha^{-1} yr^{-1}. However, the authors noted that SOC accumulation by cover crops varied depending on species assemblage and functional type. Cover crop mixtures resulted in higher SOC increases compared to mono-species cover crops, and using legumes caused more SOC increases than grass specie. Using a short-term study involving soybean-based cropping systems, Hok et al. (2015) observed a high rate (7.29 to 7.32 Mg C ha^{-1} $year^{-1}$) of SOC sequestration because of a diverse cropping system (millet/ soybean + stylosanthes guianensis and millet/ maize + brachiaria ruziziensis) was adopted in which the biomass-C inputs accrued the highest level of SOC and soil quality. In this case, the low C input of soybean was compensated by high below- and above-ground addition of millet and brachiaria; and the legume stylosanthes that had biological N fixation resulting in more balanced C and N input (Amado et al., 2006). Indeed, the positive effect of cropping systems, which included legume cover crops, on C accumulation has been reported in several studies (Boddey et al., 2010; Veloso et al., 2018). Part of this effect may be explained by the higher input of N by legumes, reducing the priming effect by SOM mining, as discussed earlier. In addition, greater diversity of plant species and the introduction of cover crops may result in better utilization of residues by microorganisms. Higher quality residues meet the stoichiometry of microorganisms and, therefore, are used more efficiently, resulting in more production of microbial by-products, which are stabilized by chemical bonding with the soil minerals (Cotrufo et al., 2013).

As seen previously, the physical protection of SOM provided by aggregates is fundamental for its accumulation in agricultural soils, playing cover crops a decisive role in the formation of stable aggregates by several factors. First, aboveground residues buffer the impact of raindrops that cause macroaggregate disruption and particle dispersion. In addition, root exudates (e.g., polysaccharides) and by-products of root-associated fungal mycelium (e.g., glomalin) act as cementing agents that bond soil particles and microaggregates. Furthermore, the acquisition of water by the roots increases the wetting-drying cycle frequency. Finally, fine roots and hyphae promote the entanglement of microaggregates to stabilize them in macroaggregates (Tisdall and Oades, 1982; Wu et al., 2014).

No-Till

No-till was initially implemented as a technique to reduce soil erosion, although several other benefits have been demonstrated using this strategy. When properly conducted, no-till increments soil structure, water infiltration, moisture conservation, nutrient cycling, and soil biological activity (Babujia et al., 2010), in addition to promoting higher crop yield (Pittelkow et al., 2015) and SOM accumulation.

The build-up of SOM by no-till occurs mainly by preserving the soil structure, which is impaired when using conventional tillage (e.g., moldboard plow, inversion tillage). Tillage operation mixes fresh residues into the soil and breaks up macroaggregates, exposing labile SOM. This high offer of a labile organic material and the increased soil aeration boost microbial activity that quickly degrades the originally physically aggregate-protected labile SOM (Briedis et al., 2018). On the other hand, no-till preserves macroaggregates, which is the way for SOM accumulation in this conservation system (De Oliveira Ferreira et al., 2018b). The limited access of labile SOM by microorganisms and their enzymes, through spatial separation and low oxygenation within the macroaggregates, decrease the mineralization of these simple structure compounds (Briedis et al., 2018; Sá et al., 2014). The continuous protection provided by the macroaggregates allows the necessary time for the labile SOM to be incorporated within microaggregates for long-term SOM stabilization in the sequence (Totsche et al., 2018). While physical protection by macroaggregates is the initial path to long-term SOM accumulation, they remain sensitive to management (Tivet et al., 2013b). Therefore, management practices should prioritize both the absence of soil disturbance, even if periodically (Inagaki et al., 2021), as well as the maintenance of permanent soil cover and the continuous supply of C-biomass (shoots and roots) that will serve as cementing agents to stabilize the macroaggregates (De Oliveira Ferreira et al., 2018b; Tisdall and Oades, 1982; Tivet et al., 2013b).

Although many results are positive, SOM accumulation under no-tillage in a deep soil profile is still controversial. Many studies have shown an effective SOM increase when no-till is used, especially in the top layers, but others claim that no-till only alters the SOM distribution in the profile (Powlson et al., 2014). It is worth noting, however, that most of the studies that show less SOM accumulation in the subsurface with no-till come from temperate regions, and some do not use all the principles of the no-till system. No-till system must be sustained by the three pillars of conservation agriculture: minimum soil disturbance, permanent soil coverage, and crop rotation with a high and diverse input of plant biomass (Reicosky, 2015). In this sense, Luo et al. (2010) in a global meta-analysis with 69 paired experiments showed that, in general, no-till accumulated more SOM than conventional tillage in the 0–30 cm layer, but differences were not found when deeper soil layers (>40 cm depth) were included. However, in the same study, the authors observed that crop frequency associated with tillage methods affected the final result, where no-till with double cropping systems (two crops per year) provided 11% more SOM than conventional tillage, even considering the 60 cm depth.

These results demonstrates that only sowing without disturbance is not enough to promote SOM accumulating in no-till, being also crucial to maintain the soil cover and, in particular, implement a crop rotation that allows a strategic combination of quantity, quality, and frequency of biomass-C additions (Sá et al., 2017). Recently, De Oliveira Ferreira et al. (2012) reported a minimum amount of biomass input of 3.21 Mg C ha^{-1} yr^{-1} (7.13 Mg ha^{-1} yr^{-1} of plant biomass) required to maintain SOM. Sá et al. (2015) reported a biomass-C input of 4.15 Mg ha^{-1} yr^{-1} (9.23 Mg ha^{-1} yr^{-1} of plant biomass) under a long-term (20 yr) no-till system to maintain the steady state of SOM. However, the cropping systems should be designed to reach values above the minimum amount of crop residues required to maintain the steady state of SOM, enhancing soil health. In tropical and subtropical regions, De Oliveira Ferreira et al. (2021) estimate that it would be necessary to add 50-75% more crop residues to achieve the benefits of long-term no-till (the maintenance phase), which is characterized by a high nutrient cycling, high photosynthesis rate of all crops, high crop residue input and SOM stabilization by macroaggregates buildup and increased faunal and soil microbial biomass carbon. In addition, it is noteworthy to highlight the importance of the roots for the SOM build-up in deep soil layers under no-till. Crop rotation should include deep-rooted species, and efficient control of toxic Al in sub-soil is also necessary to facilitate the deep penetration of roots (Briedis et al., 2012a). Therefore, when the basic premises of the no-till system are entirely followed, SOM accumulation occurs both on the top layers and subsoil (Boddey et al., 2010; Miranda et al., 2016; Sá et al., 2015; Sá et al., 2014; Sisti et al., 2004; Veloso et al., 2018).

Conclusion

Increasing SOM accumulation in agricultural soils is a current and prominent issue, as it is associated with food security and mitigation of the effects of climate change. Various mechanisms that drive SOM accumulation have been reviewed in this chapter, emphasizing that, in agricultural areas, the emphasis should be on proper management practices. In general, when the management is based on the principles of conservation agriculture (i.e., minimum soil disturbance, permanent coverage of the soil surface, and crop rotation with a high and diverse input of biomass), together with balanced fertilization (mineral and or organic) and liming acid soil, the SOM accretion is a reachable goal. However, the effectiveness of such practices and the magnitude of the SOM changes can be somewhat variable in each situation, depending to a large extent on the regional climate and the type of soil on the farm.

References

Adetunji, A.T., Ncube, B., Mulidzi, R., Lewu, F.B., 2020. Management impact and benefit of cover crops on soil quality: A review. *Soil Till. Res.* 204, 104717.

Amado, T.J.C., Bayer, C., Conceição, P.C., Spagnollo, E., de Campos, B.-H.C., da Veiga, M., 2006. Potential of Carbon Accumulation in No-Till Soils with Intensive Use and Cover Crops in Southern Brazil. *J. Environ. Qual.* 35, 1599-1607.

Angst, G., Mueller, K.E., Nierop, K.G.J., Simpson, M.J., 2021. Plant- or microbial-derived? A review on the molecular composition of stabilized soil organic matter. *Soil Biol. Biochem.* 156, 108189.

Assmann, J.M., Anghinoni, I., Martins, A.P., de Andrade, S.E.V.G., Cecagno, D., Carlos, F.S., de Faccio Carvalho, P.C., 2014. Soil carbon and nitrogen stocks and fractions in a long-term integrated crop–livestock system under no-tillage in southern Brazil. *Agric. Ecosyst. Environ.* 190, 52-59.

Babujia, L.C., Hungria, M., Franchini, J.C., Brookes, P.C., 2010. Microbial biomass and activity at various soil depths in a Brazilian oxisol after two decades of no-tillage and conventional tillage. *Soil Biol. Biochem.* 42, 2174-2181.

Bieluczyk, W., Piccolo, M.d.C., Pereira, M.G., Moraes, M.T.d., Soltangheisi, A., Bernardi, A.C.d.C., Pezzopane, J.R.M., Oliveira, P.P.A., Moreira, M.Z., Camargo, P.B.d., Dias, C.T.d.S., Batista, I., Cherubin, M.R., 2020. Integrated farming systems influence soil organic matter dynamics in southeastern Brazil. *Geoderma* 371, 114368.

Boddey, R.M., Jantalia, C.P., Conceição, P.C., Zanatta, J.A., Bayer, C., Mielniczuk, J., Dieckow, J., Dos Santos, H.P., Denardin, J.E., Aita, C., Giacomini, S.J., Alves, B.J.R., Urquiaga, S., 2010. Carbon accumulation at depth in Ferralsols under zero-till subtropical agriculture. *Glob. Change Biol.* 16, 784-795.

Bolinder, M.A., Crotty, F., Elsen, A., Frac, M., Kismányoky, T., Lipiec, J., Tits, M., Tóth, Z., Kätterer, T., 2020. The effect of crop residues, cover crops, manures and nitrogen fertilization on soil organic carbon changes in agroecosystems: a synthesis of reviews. *Mitigation and Adaptation Strategies for Global Change* 25, 929-952.

Brewer, K.M., Gaudin, A.C.M., 2020. Potential of crop-livestock integration to enhance carbon sequestration and agroecosystem functioning in semi-arid croplands. *Soil Biol. Biochem.* 149, 107936.

Briedis, C., Baldock, J., Sa, J.C.D., dos Santos, J.B., McGowan, J., Milori, D.M.B.P., 2021. Organic carbon pools and organic matter chemical composition in response to different land uses in southern Brazil. *Eur. J. Soil Sci.* 72, 1083-1100.

Briedis, C., de Moraes Sá, J.C., Lal, R., Tivet, F., de Oliveira Ferreira, A., Franchini, J.C., Schimiguel, R., da Cruz Hartman, D., Santos, J.Z.d., 2016. Can highly weathered soils under conservation agriculture be C saturated? *Catena* 147, 638-649.

Briedis, C., de Moraes Sá, J.C., Lal, R., Tivet, F., Franchini, J.C., de Oliveira Ferreira, A., da Cruz Hartman, D., Schimiguel, R., Bressan, P.T., Inagaki, T.M., Romaniw, J., Gonçalves, D.R.P., 2018. How does no-till deliver carbon stabilization and saturation in highly weathered soils? *Catena* 163, 13-23.

Briedis, C., Sá, J.C.d.M., Caires, E.F., Navarro, J.d.F., Inagaki, T.M., Boer, A., Ferreira, A.d.O., Neto, C.Q., Canalli, L.B., Bürkner dos Santos, J., 2012a. Changes in Organic Matter Pools and Increases in Carbon Sequestration in Response to Surface Liming in an Oxisol under Long-Term No-Till. *Soil Sci. Soc. Am. J.* 76, 151-160.

Briedis, C., Sá, J.C.d.M., Caires, E.F., Navarro, J.d.F., Inagaki, T.M., Boer, A., Neto, C.Q., Ferreira, A.d.O., Canalli, L.B., Santos, J.B.d., 2012b. Soil organic matter pools and carbon-protection mechanisms in aggregate classes influenced by surface liming in a no-till system. *Geoderma* 170, 80-88.

Burke, I.C., Yonker, C.M., Parton, W.J., Cole, C.V., Flach, K., Schimel, D.S., 1989. Texture, Climate, and Cultivation Effects on Soil Organic Matter Content in U.S. Grassland Soils. *Soil Sci. Soc. Am. J.* 53, 800-805.

Caires, E.F., Corrêa, J.C.L., Churka, S., Barth, G., Garbuio, F.J., 2006. Surface application of lime ameliorates subsoil acidity and improves root growth and yield of wheat in an acid soil under no-till system. *Sci. Agric.* 63, 502-509.

Carvalho, J.L.N., Raucci, G.S., Cerri, C.E.P., Bernoux, M., Feigl, B.J., Wruck, F.J., Cerri, C.C., 2010. Impact of pasture, agriculture and crop-livestock systems on soil C stocks in Brazil. *Soil Till. Res.* 110, 175-186.

Chen, R., Senbayram, M., Blagodatsky, S., Myachina, O., Dittert, K., Lin, X., Blagodatskaya, E., Kuzyakov, Y., 2014. Soil C and N availability determine the priming effect: microbial N mining and stoichiometric decomposition theories. *Glob. Change Biol.* 20, 2356-2367.

Chen, S., Zou, J., Hu, Z., Lu, Y., 2020. Temporal and spatial variations in the mean residence time of soil organic carbon and their relationship with climatic, soil and vegetation drivers. *Global Planet. Change* 195, 103359.

Chenu, C., Angers, D.A., Barré, P., Derrien, D., Arrouays, D., Balesdent, J., 2019. Increasing organic stocks in agricultural soils: Knowledge gaps and potential innovations. *Soil Till. Res.* 188, 41-52.

Coonan, E.C., Kirkby, C.A., Kirkegaard, J.A., Amidy, M.R., Strong, C.L., Richardson, A.E., 2020. Microorganisms and nutrient stoichiometry as mediators of soil organic matter dynamics. *Nutrient Cycling in Agroecosystems* 117, 273-298.

Cotrufo, M.F., Soong, J.L., Horton, A.J., Campbell, E.E., Haddix, M.L., Wall, D.H., Parton, W.J., 2015. Formation of soil organic matter via biochemical and physical pathways of litter mass loss. *Nature Geosci* 8, 776-779.

Cotrufo, M.F., Wallenstein, M.D., Boot, C.M., Denef, K., Paul, E., 2013. The Microbial Efficiency-Matrix Stabilization (MEMS) framework integrates plant litter decomposition with soil organic matter stabilization: do labile plant inputs form stable soil organic matter? *Glob. Change Biol.* 19, 988-995.

De Moraes, A., Carvalho, P.C.d.F., Anghinoni, I., Lustosa, S.B.C., Costa, S.E.V.G.d.A., Kunrath, T.R., 2014. Integrated crop–livestock systems in the Brazilian subtropics. *Eur. J. Agron.* 57, 4-9.

De Moraes Sá, J.C., Lal, R., Briedis, C., de Oliveira Ferreira, A., Tivet, F., Inagaki, T.M., Potma Gonçalves, D.R., Canalli, L.B., Burkner dos Santos, J., Romaniw, J., 2022. Can C-budget of natural capital be restored through conservation agriculture in a tropical and subtropical environment? *Environ. Pollut.* 298, 118817.

De Oliveira Ferreira, A., Amado, T.J.C., Rice, C.W., Ruiz Diaz, D.A., Briedis, C., Inagaki, T.M., Gonçalves, D.R.P., 2018a. Driving factors of soil carbon accumulation in Oxisols in long-term no-till systems of South Brazil. *Sci. Total Environ.* 622-623, 735-742.

De Oliveira Ferreira, A., De Moraes Sá, J.C., Lal, R., Jorge Carneiro Amado, T., Massao Inagaki, T., Briedis, C., Tivet, F., 2021. Can no-till restore soil organic carbon to levels under natural vegetation in a subtropical and tropical Typic Quartzipisamment? *Land Degrad. Dev.* 32, 1742-1750.

De Oliveira Ferreira, A., De Moraes Sá, J.C., Lal, R., Tivet, F., Briedis, C., Inagaki, T.M., Gonçalves, D.R.P., Romaniw, J., 2018b. Macroaggregation and soil organic carbon restoration in a highly weathered Brazilian Oxisol after two decades under no-till. *Sci. Total Environ.* 621, 1559-1567.

De Oliveira Ferreira, A., Sá, J.C.d.M., Harms, M.G., Miara, S., Briedis, C., Quadros Netto, C., Santos, J.B.d., Canalli, L.B., 2012. Carbon balance and crop residue management in dynamic equilibrium under a no-till system in Campos Gerais. *Rev. Bras. Ciênc. Solo* 36, 1583-1590.

De Stefano, A., Jacobson, M.G., 2018. Soil carbon sequestration in agroforestry systems: a meta-analysis. *Agroforestry Systems* 92, 285-299.

Dieckow, J., Bayer, C., Conceição, P.C., Zanatta, J.A., Martin-Neto, L., Milori, D.B.M., Salton, J.C., Macedo, M.M., Mielniczuk, J., Hernani, L.C., 2009. Land use, tillage, texture and organic matter stock and composition in tropical and subtropical Brazilian soils. *Eur. J. Soil Sci.* 60, 240-249.

Dungait, J.A.J., Hopkins, D.W., Gregory, A.S., Whitmore, A.P., 2012. Soil organic matter turnover is governed by accessibility not recalcitrance. *Glob. Change Biol.* 18, 1781-1796.

Epstein, H.E., Burke, I.C., Lauenroth, W.K., 2002. Regional patterns of decomposition and primary production rates in the US Great Plains. *Ecology* 83, 320-327.

Feng, W., Plante, A.F., Six, J., 2013. Improving estimates of maximal organic carbon stabilization by fine soil particles. *Biogeochemistry* 112, 81-93.

Friedlingstein, P., O'Sullivan, M., Jones, M.W., Andrew, R.M., Hauck, J., Olsen, A., Peters, G.P., Peters, W., Pongratz, J., Sitch, S., Le Quéré, C., Canadell, J.G., Ciais, P., Jackson, R.B., Alin, S., Aragão, L.E.O.C., Arneth, A., Arora, V., Bates, N.R., Becker, M., Benoit-Cattin, A., Bittig, H.C., Bopp, L., Bultan, S., Chandra, N., Chevallier, F., Chini, L.P., Evans, W., Florentie, L., Forster, P.M., Gasser, T., Gehlen, M., Gilfillan, D., Gkritzalis, T., Gregor, L., Gruber, N., Harris, I., Hartung, K., Haverd, V., Houghton, R.A., Ilyina, T., Jain, A.K., Joetzjer, E., Kadono, K., Kato, E., Kitidis, V., Korsbakken, J.I., Landschützer, P., Lefèvre, N., Lenton, A., Lienert, S., Liu, Z., Lombardozzi, D., Marland, G., Metzl, N., Munro, D.R., Nabel, J.E.M.S., Nakaoka, S.I., Niwa, Y., O'Brien, K., Ono, T., Palmer, P.I., Pierrot, D., Poulter, B., Resplandy, L., Robertson, E., Rödenbeck, C., Schwinger, J., Séférian, R., Skjelvan, I., Smith, A.J.P., Sutton, A.J., Tanhua, T., Tans, P.P., Tian, H., Tilbrook, B., van der Werf, G., Vuichard, N., Walker, A.P., Wanninkhof, R., Watson, A.J., Willis, D., Wiltshire, A.J., Yuan, W., Yue, X., Zaehle, S., 2020. Global Carbon Budget 2020. *Earth Syst. Sci. Data* 12, 3269-3340.

Garbelini, L.G., Debiasi, H., Junior, A.A.B., Franchini, J.C., Coelho, A.E., Telles, T.S., 2022. Diversified crop rotations increase the yield and economic efficiency of grain production systems. *European Journal of Agronomy* 137, 126528.

Georgiou, K., Jackson, R.B., Vindušková, O., Abramoff, R.Z., Ahlström, A., Feng, W., Harden, J.W., Pellegrini, A.F.A., Polley, H.W., Soong, J.L., Riley, W.J., Torn, M.S., 2022. Global stocks and capacity of mineral-associated soil organic carbon. *Nat. Commun.* 13, 3797.

Gonçalves, D.R.P., Sá, J.C.d.M., Mishra, U., Fornari, A.J., Furlan, F.J.F., Ferreira, L.A., Inagaki, T.M., Romaniw, J., de Oliveira Ferreira, A., Briedis, C., 2019. Conservation agriculture based on diversified and high-performance production system leads to soil carbon sequestration in subtropical environments. *Journal of Cleaner Production* 219, 136-147.

Gross, A., Bromm, T., Glaser, B., 2021. Soil organic carbon sequestration after biochar application: A global meta-analysis. *Agronomy* 11, 2474.

Han, L., Sun, K., Jin, J., Xing, B., 2016a. Some concepts of soil organic carbon characteristics and mineral interaction from a review of literature. *Soil Biol. Biochem.* 94, 107-121.

Han, P., Zhang, W., Wang, G., Sun, W., Huang, Y., 2016b. Changes in soil organic carbon in croplands subjected to fertilizer management: a global meta-analysis. *Scientific reports* 6, 1-13.

Hassink, J., 1997. The capacity of soils to preserve organic C and N by their association with clay and silt particles. *Plant Soil* 191, 77-87.

Hok, L., Sá, J.C.d.M., Boulakia, S., Reyes, M., Leng, V., Kong, R., Tivet, F.E., Briedis, C., Hartman, D., Ferreira, L.A., Magno, T., Pheav, S., 2015. Short-term conservation agriculture and biomass-C input impacts on soil C dynamics in a savanna ecosystem in Cambodia. *Agric. Ecosyst. Environ.* 214, 54-67.

Inagaki, T.M., Sá, J.C.d.M., Tormena, C.A., Dranski, A., Muchalak, A., Briedis, C., de Oliveira Ferreira, A., Giarola, N.F.B., da Silva, Á.P., 2021. Mechanical and biological chiseling impacts on soil organic C stocks, root growth, and crop yield in a long-term no-till system. *Soil Till. Res.* 211, 104993.

Jian, J., Du, X., Reiter, M.S., Stewart, R.D., 2020. A meta-analysis of global cropland soil carbon changes due to cover cropping. *Soil Biol. Biochem.* 143, 107735.

Johnston, A.M., Bruulsema, T.W., 2014. 4R nutrient stewardship for improved nutrient use efficiency. *Procedia Engineering 83*, 365-370.

Jose, S., 2009. Agroforestry for ecosystem services and environmental benefits: an overview. *Agroforestry systems* 76, 1-10.

Kaiser, K., Kalbitz, K., 2012. Cycling downwards – dissolved organic matter in soils. *Soil Biol. Biochem.* 52, 29-32.

Kirkby, C.A., Richardson, A.E., Wade, L.J., Batten, G.D., Blanchard, C., Kirkegaard, J.A., 2013. Carbon-nutrient stoichiometry to increase soil carbon sequestration. *Soil Biol. Biochem.* 60, 77-86.

Kirkby, C.A., Richardson, A.E., Wade, L.J., Passioura, J.B., Batten, G.D., Blanchard, C., Kirkegaard, J.A., 2014. Nutrient availability limits carbon sequestration in arable soils. *Soil Biol. Biochem.* 68, 402-409.

Kleber, M., Eusterhues, K., Keiluweit, M., Mikutta, C., Mikutta, R., Nico, P.S., 2015. Mineral–organic associations: formation, properties, and relevance in soil environments. *Advances in agronomy* 130, 1-140.

Kögel-Knabner, I., Guggenberger, G., Kleber, M., Kandeler, E., Kalbitz, K., Scheu, S., Eusterhues, K., Leinweber, P., 2008. Organo-mineral associations in temperate soils: Integrating biology, mineralogy, and organic matter chemistry. *J. Plant Nutr. Soil Sci.* 171, 61-82.

Kravchenko, A.N., Guber, A.K., 2017. Soil pores and their contributions to soil carbon processes. *Geoderma* 287, 31-39.

Krull, E.S., Baldock, J.A., Skjemstad, J.O., 2003. Importance of mechanisms and processes of the stabilisation of soil organic matter for modelling carbon turnover. *Funct. Plant Biol.* 30, 207-222.

Lal, R., 2004. Soil carbon sequestration to mitigate climate change. Geoderma 123, 1-22.

Lal, R., 2016. Soil health and carbon management. *Food and Energy Security* 5, 212-222.

Lal, R., 2020. Soil organic matter content and crop yield. *J. Soil Water Conserv.* 75, 27A.

Liang, C., Amelung, W., Lehmann, J., Kästner, M., 2019. Quantitative assessment of microbial necromass contribution to soil organic matter. *Glob. Change Biol.* 25, 3578-3590.

Liang, C., Schimel, J.P., Jastrow, J.D., 2017. The importance of anabolism in microbial control over soil carbon storage. *Nature microbiology* 2, 1-6.

Lehmann, J., Abiven, S., Kleber, M., Pan, G., Singh, B.P., Sohi, S.P., Zimmerman, A.R., 2015. Persistence of biochar in soil. *Biochar for environmental management.* Routledge, pp. 267-314.

Lehmann, J., Kern, D., German, L., McCann, J., Martins, G.C., Moreira, A., 2003. Soil fertility and production potential. *Amazonian dark earths.* Springer, pp. 105-124.

Loganathan, N., Ferguson, B.O., Arey, B., Argersinger, H.E., Bowers, G.M., 2020. A Mechanistic Exploration of Natural Organic Matter Aggregation and Surface Complexation in Smectite Mesopores. *The Journal of Physical Chemistry A* 124, 9832-9843.

Lorenz, K., Lal, R., 2014a. Biochar application to soil for climate change mitigation by soil organic carbon sequestration. *J. Plant Nutr. Soil Sci.* 177, 651-670.

Lorenz, K., Lal, R., 2014b. Soil organic carbon sequestration in agroforestry systems. A review. *Agronomy for Sustainable Development* 34, 443-454.

Lu, X., 2020. Fertilizer Types Affect Soil Organic Carbon Content and Crop Production: A Meta-analysis. *Agricultural Research* 9, 94-101.

Luo, Z., Viscarra-Rossel, R.A., Qian, T., 2021. Similar importance of edaphic and climatic factors for controlling soil organic carbon stocks of the world. *Biogeosciences* 18, 2063-2073.

Luo, Z., Wang, E., Sun, O.J., 2010. Can no-tillage stimulate carbon sequestration in agricultural soils? A meta-analysis of paired experiments. *Agric. Ecosyst. Environ.* 139, 224-231.

Majumder, S., Neogi, S., Dutta, T., Powel, M.A., Banik, P., 2019. The impact of biochar on soil carbon sequestration: meta-analytical approach to evaluating environmental and economic advantages. *J. Environ. Manage.* 250, 109466.

Manzoni, S., Taylor, P., Richter, A., Porporato, A., Ågren, G.I., 2012. Environmental and stoichiometric controls on microbial carbon-use efficiency in soils. *New Phytol.* 196, 79-91.

Marschner, B., Brodowski, S., Dreves, A., Gleixner, G., Gude, A., Grootes, P.M., Hamer, U., Heim, A., Jandl, G., Ji, R., Kaiser, K., Kalbitz, K., Kramer, C., Leinweber, P., Rethemeyer, J., Schäffer, A., Schmidt, M.W.I., Schwark, L., Wiesenberg, G.L.B., 2008. How relevant is recalcitrance for the stabilization of organic matter in soils? *J. Plant Nutr. Soil Sci.* 171, 91-110.

McClelland, S.C., Paustian, K., Schipanski, M.E., 2021. Management of cover crops in temperate climates influences soil organic carbon stocks: a meta-analysis. *Ecol. Appl.* 31, e02278.

Mikutta, R., Kleber, M., Torn, M., Jahn, R., 2006. Stabilization of Soil Organic Matter: Association with Minerals or Chemical Recalcitrance? *Biogeochemistry* 77, 25-56.

Miltner, A., Bombach, P., Schmidt-Brücken, B., Kästner, M., 2012. SOM genesis: microbial biomass as a significant source. *Biogeochemistry* 111, 41-55.

Miranda, E., Carmo, J., Couto, E., Camargo, P., 2016. Long-Term Changes in Soil Carbon Stocks in the Brazilian Cerrado Under Commercial Soybean. *Land Degrad. Dev.* 27, 1586-1594.

Muchane, M.N., Sileshi, G.W., Gripenberg, S., Jonsson, M., Pumarino, L., Barrios, E., 2020. Agroforestry boosts soil health in the humid and sub-humid tropics: A meta-analysis. *Agric. Ecosyst. Environ.* 295, 106899.

Nearing, M.A., Jetten, V., Baffaut, C., Cerdan, O., Couturier, A., Hernandez, M., Le Bissonnais, Y., Nichols, M.H., Nunes, J.P., Renschler, C.S., Souchère, V., van Oost, K., 2005. Modeling response of soil erosion and runoff to changes in precipitation and cover. *Catena* 61, 131-154.

Nicoloso, R.S., Rice, C.W., Amado, T.J.C., Costa, C.N., Akley, E.K., 2018. Carbon saturation and translocation in a no-till soil under organic amendments. *Agric. Ecosyst. Environ.* 264, 73-84.

Nora, D.D., Amado, T.J.C., 2013. Improvement in Chemical Attributes of Oxisol Subsoil and Crop Yields under No-Till. *Agron. J.,* 1393-1403.

Paul, E.A., Harris, D., Collins, H.P., Schulthess, U., Robertson, G.P., 1999. Evolution of CO2 and soil carbon dynamics in biologically managed, row-crop agroecosystems. *Appl. Soil Ecol.* 11, 53-65.

Perkons, U., Kautz, T., Uteau, D., Peth, S., Geier, V., Thomas, K., Lütke Holz, K., Athmann, M., Pude, R., Köpke, U., 2014. Root-length densities of various annual crops following crops with contrasting root systems. *Soil Till.* Res. 137, 50-57.

Pittelkow, C.M., Linquist, B.A., Lundy, M.E., Liang, X., van Groenigen, K.J., Lee, J., van Gestel, N., Six, J., Venterea, R.T., van Kessel, C., 2015. When does no-till yield more? A global meta-analysis. *Field Crops Res.* 183, 156-168.

Plaza, C., Courtier-Murias, D., Fernández, J.M., Polo, A., Simpson, A.J., 2013. Physical, chemical, and biochemical mechanisms of soil organic matter stabilization under conservation tillage systems: A central role for microbes and microbial by-products in C sequestration. *Soil Biol. Biochem.* 57, 124-134.

Plaza, C., Gascó, G., Méndez, A.M., Zaccone, C., Maestre, F.T., 2018. Chapter 2 - Soil Organic Matter in Dryland Ecosystems. In: Garcia, C., Nannipieri, P., Hernandez, T. (Eds.), *The Future of Soil Carbon.* Academic Press, pp. 39-70.

Poeplau, C., Don, A., 2015. Carbon sequestration in agricultural soils via cultivation of cover crops – A meta-analysis. *Agric. Ecosyst. Environ.* 200, 33-41.

Poeplau, C., Helfrich, M., Dechow, R., Szoboszlay, M., Tebbe, C.C., Don, A., Greiner, B., Zopf, D., Thumm, U., Korevaar, H., 2019. Increased microbial anabolism contributes to soil carbon sequestration by mineral fertilization in temperate grasslands. *Soil Biol. Biochem.* 130, 167-176.

Poeplau, C., Zopf, D., Greiner, B., Geerts, R., Korvaar, H., Thumm, U., Don, A., Heidkamp, A., Flessa, H., 2018. Why does mineral fertilization increase soil carbon stocks in temperate grasslands? *Agric. Ecosyst. Environ.* 265, 144-155.

Poirier, V., Basile-Doelsch, I., Balesdent, J., Borschneck, D., Whalen, J.K., Angers, D.A., 2020. Organo-mineral interactions are more important for organic matter retention in subsoil than topsoil. *Soil Syst.* 4, 4.

Powlson, D.S., Stirling, C.M., Jat, M.L., Gerard, B.G., Palm, C.A., Sanchez, P.A., Cassman, K.G., 2014. Limited potential of no-till agriculture for climate change mitigation. *Nature Climate Change* 4, 678-683.

Puget, P., Chenu, C., Balesdent, J., 2000. Dynamics of soil organic matter associated with particle-size fractions of water-stable aggregates. *Eur. J. Soil Sci.* 51, 595-605.

Quilty, J.R., Cattle, S.R., 2011. Use and understanding of organic amendments in Australian agriculture: a review. *Soil Research* 49, 1-26.

Rasse, D., Rumpel, C., Dignac, M.-F., 2005. Is soil carbon mostly root carbon? Mechanisms for a specific stabilisation. *Plant Soil* 269, 341-356.

Reicosky, D.C., 2015. Conservation tillage is not conservation agriculture. *J. Soil Water Conserv.* 70, 103A.

Reis, C.E.S.d., Dick, D.P., Caldas, J.d.S., Bayer, C., 2014. Carbon sequestration in clay and silt fractions of Brazilian soils under conventional and no-tillage systems. *Sci. Agric. 71,* 292-301.

Romaniw, J., de Moraes Sá, J.C., Lal, R., de Oliveira Ferreira, A., Inagaki, T.M., Briedis, C., Gonçalves, D.R.P., Canalli, L.B., Padilha, A., Bressan, P.T., 2021. C-offset and crop energy efficiency increase due industrial poultry waste use in long-term no-till soil minimizing environmental pollution. *Environ. Pollut.* 275, 116565.

Roth, C.H., Pavan, M.A., 1991. Effects of lime and gypsum on clay dispersion and infiltration in samples of a Brazilian Oxisol. *Geoderma* 48, 351-361.

Rowley, M.C., Grand, S., Verrecchia, É.P., 2018. Calcium-mediated stabilisation of soil organic carbon. *Biogeochemistry* 137, 27-49.

Rumpel, C., Kögel-Knabner, I., 2011. Deep soil organic matter—a key but poorly understood component of terrestrial C cycle. *Plant Soil* 338, 143-158.

Sá, J.C.d.M., Lal, R., Cerri, C.C., Lorenz, K., Hungria, M., de Faccio Carvalho, P.C., 2017. Low-carbon agriculture in South America to mitigate global climate change and advance food security. *Environ. Int.* 98, 102-112.

Sá, J.C.d.M., Séguy, L., Tivet, F., Lal, R., Bouzinac, S., Borszowskei, P.R., Briedis, C., dos Santos, J.B., da Cruz Hartman, D., Bertoloni, C.G., Rosa, J., Friedrich, T., 2015. Carbon Depletion by Plowing and its Restoration by No-Till Cropping Systems in Oxisols of Subtropical and Tropical Agro-Ecoregions in Brazil. *Land Degrad. Dev.* 26, 531-543.

Sá, J.C.d.M., Tivet, F., Lal, R., Briedis, C., Hartman, D.C., dos Santos, J.Z., dos Santos, J.B., 2014. Long-term tillage systems impacts on soil C dynamics, soil resilience and agronomic productivity of a Brazilian Oxisol. *Soil Till. Res.* 136, 38-50.

Salton, J.C., Mercante, F.M., Tomazi, M., Zanatta, J.A., Concenço, G., Silva, W.M., Retore, M., 2014. Integrated crop-livestock system in tropical Brazil: Toward a sustainable production system. *Agric. Ecosyst. Environ.* 190, 70-79.

Schimel, D.S., Braswell, B.H., Holland, E.A., McKeown, R., Ojima, D.S., Painter, T.H., Parton, W.J., Townsend, A.R., 1994. Climatic, edaphic, and biotic controls over storage and turnover of carbon in soils. *Global Biogeochem. Cy.* 8, 279-293.

Schmidt, M.W.I., Torn, M.S., Abiven, S., Dittmar, T., Guggenberger, G., Janssens, I.A., Kleber, M., Kogel-Knabner, I., Lehmann, J., Manning, D.A.C., Nannipieri, P., Rasse, D.P., Weiner, S., Trumbore, S.E., 2011. Persistence of soil organic matter as an ecosystem property. *Nature* 478, 49-56.

Scotti, R., Bonanomi, G., Scelza, R., Zoina, A., Rao, M.A., 2015. Organic amendments as sustainable tool to recovery fertility in intensive agricultural systems. *J. Soil Sci. Plant Nutr.* 15, 333-352.

Shi, L., Feng, W., Xu, J., Kuzyakov, Y., 2018. Agroforestry systems: Meta-analysis of soil carbon stocks, sequestration processes, and future potentials. *Land Degrad. Dev.* 29, 3886-3897.

Singh, M., Sarkar, B., Sarkar, S., Churchman, J., Bolan, N., Mandal, S., Menon, M., Purakayastha, T.J., Beerling, D.J., 2018. Stabilization of soil organic carbon as influenced by clay mineralogy. *Advances in agronomy* 148, 33-84.

Singh, N., Abiven, S., Torn, M.S., Schmidt, M.W.I., 2012. Fire-derived organic carbon in soil turns over on a centennial scale. *Biogeosciences* 9, 2847-2857.

Sisti, C.P.J., dos Santos, H.P., Kohhann, R., Alves, B.J.R., Urquiaga, S., Boddey, R.M., 2004. Change in carbon and nitrogen stocks in soil under 13 years of conventional or zero tillage in southern Brazil. *Soil Till. Res.* 76, 39-58.

Six, J., Conant, R.T., Paul, E.A., Paustian, K., 2002. Stabilization mechanisms of soil organic matter: Implications for C-saturation of soils. *Plant Soil* 241, 155-176.

Six, J., Elliott, E.T., Paustian, K., 1999. Aggregate and Soil Organic Matter Dynamics under Conventional and No-Tillage Systems. *Soil Sci. Soc. Am. J.* 63, 1350-1358.

Six, J., Elliott, E.T., Paustian, K., 2000. Soil macroaggregate turnover and microaggregate formation: a mechanism for C sequestration under no-tillage agriculture. *Soil Biol. Biochem.* 32, 2099-2103.

Souza, I.F., Archanjo, B.S., Hurtarte, L.C.C., Oliveros, M.E., Gouvea, C.P., Lidizio, L.R., Achete, C.A., Schaefer, C.E.R., Silva, I.R., 2017. Al-/Fe-(hydr)oxides–organic carbon associations in Oxisols — From ecosystems to submicron scales. *Catena* 154, 63-72.

Tisdall, J.M., Oades, J.M., 1982. Organic matter and water-stable aggregates in soils. *J. Soil Sci.* 33, 141-163.

Tivet, F., Sá, J.C.d.M., Lal, R., Borszowskei, P.R., Briedis, C., dos Santos, J.B., Sá, M.F.M., Hartman, D.d.C., Eurich, G., Farias, A., Bouzinac, S., Séguy, L., 2013a. Soil organic carbon fraction losses upon continuous plow-based tillage and its restoration by diverse biomass-C inputs under no-till in sub-tropical and tropical regions of Brazil. *Geoderma* 209-210, 214-225.

Tivet, F., Sá, J.C.d.M., Lal, R., Briedis, C., Borszowskei, P.R., dos Santos, J.B., Farias, A., Eurich, G., Hartman, D.d.C., Nadolny Junior, M., Bouzinac, S., Séguy, L., 2013b. Aggregate C depletion by plowing and its restoration by diverse biomass-C inputs under no-till in sub-tropical and tropical regions of Brazil. *Soil Till. Res.* 126, 203-218.

Totsche, K.U., Amelung, W., Gerzabek, M.H., Guggenberger, G., Klumpp, E., Knief, C., Lehndorff, E., Mikutta, R., Peth, S., Prechtel, A., Ray, N., Kögel-Knabner, I., 2018. Microaggregates in soils. *J. Plant Nutr. Soil Sci.* 181, 104-136.

Veloso, M.G., Angers, D.A., Tiecher, T., Giacomini, S., Dieckow, J., Bayer, C., 2018. High carbon storage in a previously degraded subtropical soil under no-tillage with legume cover crops. *Agric. Ecosyst. Environ.* 268, 15-23.

von Lützow, M., Kögel-Knabner, I., Ekschmitt, K., Matzner, E., Guggenberger, G., Marschner, B., Flessa, H., 2006. Stabilization of organic matter in temperate soils: mechanisms and their relevance under different soil conditions – a review. *Eur. J. Soil Sci.* 57, 426-445.

Wang, J., Xiong, Z., Kuzyakov, Y., 2016. Biochar stability in soil: meta-analysis of decomposition and priming effects. *Gcb Bioenergy 8*, 512-523.

Weber, K., Quicker, P., 2018. Properties of biochar. *Fuel* 217, 240-261.

Wei, X., Shao, M., Gale, W., Li, L., 2014. Global pattern of soil carbon losses due to the conversion of forests to agricultural land. *Scientific Reports* 4, 4062.

Weil, R.R., Magdoff, F., 2004. Significance of soil organic matter to soil quality and health. *Soil organic matter in sustainable agriculture,* 1-43.

Wiesmeier, M., Urbanski, L., Hobley, E., Lang, B., von Lützow, M., Marin-Spiotta, E., van Wesemael, B., Rabot, E., Ließ, M., Garcia-Franco, N., Wollschläger, U., Vogel, H.-J., Kögel-Knabner, I., 2019. Soil organic carbon storage as a key function of soils - A review of drivers and indicators at various scales. *Geoderma* 333, 149-162.

Wiseman, C.L.S., Püttmann, W., 2005. Soil organic carbon and its sorptive preservation in central Germany. *Eur. J. Soil Sci.* 56, 65-76.

Wu, Q.-S., Cao, M.-Q., Zou, Y.-N., He, X.-h., 2014. Direct and indirect effects of glomalin, mycorrhizal hyphae and roots on aggregate stability in rhizosphere of trifoliate orange. *Scientific reports* 4, 1-8.

Zhou, Y., Qin, S., Verma, S., Sar, T., Sarsaiya, S., Ravindran, B., Liu, T., Sindhu, R., Patel, A.K., Binod, P., 2021. Production and beneficial impact of biochar for environmental application: A comprehensive review. *Bioresour. Technol.* 337, 125451.

Zhu, X., Liu, W., Chen, J., Bruijnzeel, L.A., Mao, Z., Yang, X., Cardinael, R., Meng, F.-R., Sidle, R.C., Seitz, S., Nair, V.D., Nanko, K., Zou, X., Chen, C., Jiang, X.J., 2020. Reductions in water, soil and nutrient losses and pesticide pollution in agroforestry practices: a review of evidence and processes. *Plant Soil* 453, 45-86.

Zinn, Y.L., Lal, R., Bigham, J.M., Resck, D.V.S., 2007. Edaphic Controls on Soil Organic Carbon Retention in the Brazilian Cerrado: Texture and Mineralogy. *Soil Sci. Soc. Am. J.* 71, 1204-1214.

Chapter 6

The Microbiological Evaluation of Agricultural Lands

Muazzez Gürgan[1,*]
and Sevinç Yeşilyurt[2]

[1]Department of Biology, Tekirdağ Namık Kemal University, Tekirdag, Turkey
[2]Department of Soil Science and Plant Nutrition, Tekirdağ Namık Kemal University, Tekirdağ, Turkey

Abstract

The biological activity of soils is one of the most important parameters in sustainable agriculture and is crucial for the transformation of plant nutrients into extractable forms. Bacteria are the most important group of microorganisms that contribute to the vitality of agricultural soils. Bacteria, the largest group of microorganisms at neutral pH, increase the solubility and extractability of plant nutrients. They also help to improve the physical and chemical properties of the soil and create an environment suitable for biological production. However, due to some agricultural systems, the biological properties of the soil decrease over time. Bio-fertilizers step into the field to eliminate this problem. Bio-fertilizers are environmentally friendly organic agricultural inputs. Enrichment of agricultural soils with nitrogen results in enrichment of soils with phosphorus due to increased solubility of phosphate compounds and also stimulates the secretion of some chemicals that promote plant development. Biofertilizers also reduce the use of chemical fertilizers and protect against pathogenic fungi. Rhizobium, Azosipirillum, Azotobacter, Bacillus, Pseudomonas are among the most important groups of organisms. Nevertheless, bioremediation becomes

---

* Corresponding Author's Email: mgurgan@nku.edu.tr.

In: Agricultural Soils
Editors: R. Nogueira de Sousa, T. Nascimento Pessoa and K. Ferreira Mendes
ISBN: 979-8-89113-000-5

prominent in the natural alleviation of micro-nutrient toxicity and especially heavy metal pollution resulting from industry, traffic, fertilizers, pesticides, municipal wastes, etc. This chapter focuses on the relationship of biological characteristics of agricultural lands with the environment and solutions.

Keywords: agricultural land, soil biology, bacteria, bioremediation

Introduction

The expression of agricultural problems nowadays depends on experts' comprehension level of the basic rules of plant and soil science and their correct evaluation of the results they obtain. The most important step is the productivity analyses of soil and plant and accurate evaluation of them (Müftüoğlu et al., 2012).

Sustainable agricultural production depends on the content and availability of sufficient and balanced plant essential nutrient elements in the soil which is the production site. Instead of expressing how long the cultivation can be carried out in the soil without the addition of plant nutrients to the soil, sustainability should be maintained by the necessary addition. Techniques to determine the soil potential of production provide the determination of how much, when, and by which methods the nutrients should be added to the agricultural soils. That is why appropriate and cost-effective methods should be developed to determine the insufficient nutrient elements in soil (Adiloğlu and Sağlam 2015; Koca et al., 2019).

The yield increase in agricultural production depends on many soil factors, especially soil structure, nutrient capacity, and textural fractions are important. Soil chemical analysis results therefore will suggest the amount the fertilizer to be added depending on the class and kind of soil and especially how much the available levels of essential elements are in that soil. Another important point about the availability of nutrient elements is the different nutrient needs of plants. Plants with high yield potential require a high amount of available essential elements in the soil. This situation becomes significant when nitrogen, a macronutrient element, is applied in high concentrations (Adiloğlu and Adiloğlu 2005).

Soil organic matter which forms the soil's nutrition capacity, and the basis of sustainable productivity is one of the main components of soil that directly or indirectly affect many properties of the soil (Table 1, Figure 1). Soil organic matter is equal to the sum of the matter found in different steps of deterioration of plant or animal-sourced waste material which are not the same as the fresh matter. Broken-down organic compounds give different colors to the soil from brown to black depending on the different ecological zones and the dominant cations in the mineral section. Organic soil compounds also form the source of life in the soil. The source of carbon compounds required by the heterotrophs and CO_2 needed by autotrophs are sourced from the soil organic matter. Moreover, microorganisms make their energy from the exothermic reactions that form CO_2, NO_3, and SO_4 by biological oxidation of organic matter (Marschner 2008; Mengel and Kirkby 2001; Gürgan and Adiloğlu 2021a).

Table 1. Organic matter evaluation standards

Organic matter (%)	Interpretation
0-1	Very low
1-2	Low
2-3	Sufficient
3-4	Good
> 4	Very good

Source: Adapted from Adiloğlu, 2013.

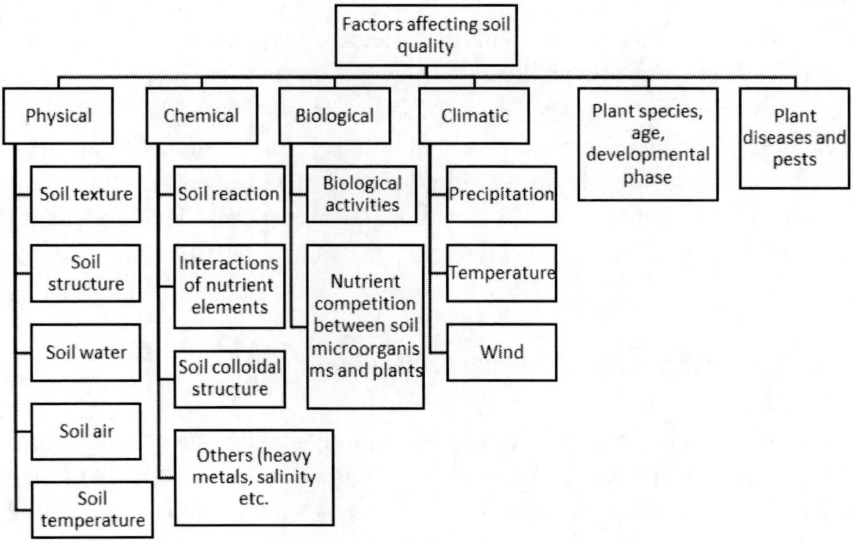

Figure 1. Factors affecting soil quality parameters.

Vitality of Soil: Soil Biology

Soil biology has gained notoriety in recent years. Protection of natural productivity power of agricultural soils, side effects of artificial compounds reaching to soils, scientific data on independent and symbiotic nitrogen fixation made researchers pay attention to healthy soil biology. Similarly, scientific data on pathogen and saprophytic microflora is closely related to plant diseases. New inventions and techniques developed in recent years in ecology, biochemistry, etc. helped the development of soil biology and made it gain a place in agricultural production, although there is still a long way to go. Soil biology, or more specific study of the soil microbiome is a complex structure that should include different scientific branches (Tok 1997; Adiloğlu 2020).

In order to maintain the sustainability of soil productivity, strategies to increase soil quality are required. One of these strategies is the increase of beneficial microorganism populations in the rhizosphere with biological fertilizers (biofertilizers). Biofertilizers such as those containing Rhizobium Azospirillum, Azotobacter, Cyanobacteria, and Azolla, and Mycorrhiza containing fertilizers applied to soil increase soil productivity and biodiversity, decrease the need for chemical fertilizers, degrade the organic wastes, and decrease the pollution caused by the chemical inputs (Adiloğlu et al., 2019; Akay and Adiloğlu 2020; Adiloğlu et al., 2021). Biofertilizers are the ideal fertilizers for sustainable agricultural systems, easy to use and economical. Such biotechnologically produced fertilizers contribute to soil productivity by fixing atmospheric nitrogen in soil and plants, increasing the solubility of insoluble phosphate compounds, and stimulating some compounds that trigger plant development. However, biofertilizers themselves cannot replace chemical inputs but can reduce their use and support sustainable agricultural systems (Karaman et al., 2012, Karaçal and Tüfenkci 2010).

Nutrition Capacity of Soils and Bacteria

There are five main groups of microorganisms in soil: bacteria, actinomycetes, fungi, algae, and protozoa. A soil ecosystem is a dynamic system that contains microorganisms and their energy source organic compounds. Bacteria are the most frequently found microorganisms in the microbial population. Moreover, the area they take, and their mass is very low. Their size is not more than a few microns. That is why they form only half of the total microorganism mass. This ratio differs in the forest, agricultural land, or herbage soils. In well-aerated soils, bacteria and fungi are the leading microorganisms (Lowenfels and Lewis, 2010). In soils of different ecological circumstances (freeze-thaw, wetting-drying, or high temperature), bacillus bacteria can survive by their endospore structures. These bacteria form endospores when the conditions change to unfavorable conditions and can resurrect to vegetative form when the conditions reach too optimum. Endospore-forming bacteria can be aerobic or anaerobic. Bacteria are rarely found in soil solution because bacteria are found mostly adhered to clay or humus particles (Sessitsch et al., 2001). Some bacteria are found in positions like colonies. Adherence of bacteria or colonies to colloidal clay or humus particles is a result of the electrostatic properties of these particles, and the interaction of bacterial cells segregates with these particles, and also because bacteria can find substrates easily this way. Hence, the free movement of bacteria is limited in the soil solution, but their biochemical activities increase in the soil solution. The diversity and number of bacteria depend on the cultural application. As the amount of root and shoot plant mixture in the soil increases, the number of bacteria also increases. This increase is most often observed in warmer climates. In short, soil organic matter and ecological conditions are the most important elements. In general, cultivated agricultural soils in a conservative system contain more bacteria than poor soils in the same region. This can be a result of more aeration of the soil.

However, when the organic matter of the cultivated soil decreases, this situation reverses. There is a rational relationship between soil bacterial number with the biological structure and activity of agriculture soil, and the climatic zone (Karaman et al., 2012). In warmer climates, the bacterial number shows a sinusoidal distribution depending on the seasons because of the freeze-thaw and wetting-drying cycles of the soil. In such regions, bacteria number is maximum in spring. The bacterial number is very low in arctic regions because of the low temperature. In these regions, soil can be frozen for months. When the temperature rises, some bacterial species may enter vegetative form after the dormant stage. Soil humidity is the limiting factor in arid and semi-arid regions. That is why in these regions, bacteria a-other than endospore-forming bacilli are rare. On the other hand, in tropical regions, the number of bacteria is always high, and the population structure is stable. The raw molecules and plant wastes required for bacterial growth are generally at optimal levels. Therefore, bacterial diversity and abundance are related to vegetative growth in the soil (Tok 1997; Arun 2007).

Nitrogen and Bacteria

Nitrogen macro element is one of the most important elements for sustainable soil productivity. Providing this element, determining where this element comes from, and which bacteria take place in the provision of this element can be revealed by the exhibition of bacteria which also mineralize the organic matter in the soil. These bacteria also take place in the bioremediation of polluted soils. They also play a role as fertilizer to fulfill the organic matter needs. They play a role in the mineralization of naturally found nitrogen sources and diversify the resulting outputs. They are one of the most important factors to make the plant nutrient elements (N, P, K, Ca, Mg, S, Fe, Cu, Zn, Mn, B, Mo, Cl, and Ni) in the plant tissues and soil particles available (Gürgan and Adiloğlu 2020; Gürgan and Adiloğlu 2021a; Adiloğlu and Adiloğlu 2021b).

On the other hand, nitrogen and carbon in plant tissues affect the degradation rate of plant tissues and hence humification and is important for plant nutrition. Nitrogen is especially important as it significantly affects the biological degradation rate in soil. Plant tissues are synthesized differently according to their nitrogen content. Plant proteins are classified as simple and complex proteins. Simple proteins such as gluten, globulins, and albumins are usually found in the structure of many grains and legumes. Lipoproteins, chromoproteins, and glycoproteins, which are among the complex proteins, have more complex protein structures; these have a slower rate of degradation. Some compounds contain nitrogen but are not part of the protein structure. Purines and pyrimidines and their derivatives are the most important complex proteins. Examples include teofiline, caffeine, kinin, papaverine, xanthin, and guanine. There are also nitrogen compounds such as amines, amides, and amino acids in various steps of nitrogen assimilation in plant tissues (Table 2).

Carbohydrate compounds differ in different plants. For example, sugars can be found in monomers (ribose, arabinose, xylose, glucose, galactose, etc.), oligomers (saccharose, cellobiose, etc.). or polymers (starch, inulin, cellulose, etc.). As the polymer content of organic tissues increases, the resistance to mineralization increases. The functional groups of sugars that are aldo or keto also affect biological degradation. Polyuronides, which are an important source of negative ions in the soil, are organic compounds such as glucose or galactose bound. There are also hydrocarbons such as paraffin, organic acids, alcohols, esters, and aldehydes in plant tissues.

Table 2. Nitrogen compounds found in plant and microbial cells

Nitrogen compounds				
Aminoacids and Proteins	Nucleic acids	Amines	Vitamins	Purines and Pyrimidines

Secondary biomass that forms organic matter in the soil, apart from higher plants, belongs to soil microbial communities and soil fauna. The nitrogen percentage in these organisms, which leave less organic matter to the soil than the higher plants, is usually higher (2.5-6.69%) and they are important elements that play a role in soil aggregation. The bacteria that take place in the nitrogen cycle are:

- Nitrog-fixing aerobic bacteria (Azotobacter chroococcunz, A. vinelandii, A. beijerinckii, A. armeniacus (Pandey and Sreelu 2006))
- Nitrogen-fixing anaerobic bacteria (e.g., Clostridium species)
- Proteolytic bacteria (e.g., Pseudomonas, Bacillus, Clostridium species)
- Ammonification bacteria e.g., Pseudomonas, Bacillus, Clostridium species)
- Nitrifying bacteria (Nitrosomonas, Nitrosococcus, Nitrobacter species)
- Denitrifying bacteria (e.g., Thiobacillus denitrificans, Micrococcus denitrificans)

The most common bacteria used in fertilizer mixtures for nitrogen input are *Rhizobium* or *Bradyrhizobium* species, which can fix atmospheric nitrogen and form symbiosis with leguminous plants. These bacteria can fix about 5-10 kg. ha^{-1} of nitrogen per year. The fertilizer inoculated with *rhizobium* species appropriate for the legume to be grown is usually mixed with the seeds of the plant. In this way, the bacteria living in the soil reproduce rapidly under optimal conditions and form nodules that start nitrogen fixation in the early development stage of the plant and meet the nitrogen needs of the plant.

Fertilizers Containing AzospirillumAzospirillum

These heterotrophic microorganisms live especially with C4 plants (Eckert et al., 2001). They can fix around 2-4 kg N/da and they form plant growth stimulators.

Microbial Fertilizers Containing Azotobacter

These bacteria are aerobic free-living bacteria that can fix atmospheric nitrogen (N_2). Although they are naturally found in neutral or alkaline soils, their number is not high (Karaman et al., 2021).

Phosphorus and Bacteria

Phosphorus is the second macronutrient required for optimum growth and development. It forms about 0.2% of the plant's dry biomass and takes numerous physiological and biochemical reactions in plants (Theodorous and Plaxton 1993).

Phosphorus is a significant element of proteins, nucleic acids, enzymes, coenzymes, and phospholipids. It is an absolute element in the synthesis of energy-rich elements such as NAD, NADH, ADP, and ATP which take place in both respiration and photosynthesis. It plays a role in many physiological events such as protein synthesis with the help of energy-rich bridges. Plants cannot reproduce, produce seeds or perform nouaison, they cannot develop and perform their normal functions. It provides energy production by the formation of glucose and starch in photosynthesis and oxidation of them in respiration. It accelerates root development, early development, flowering, seed production, and fruit production. Moreover, it takes place in the transport of nutrient elements and other compounds. Energy is especially required for the transportation of organic compounds to storage organs and seeds (Mengel and Kirkby 2001; Buehler et al., 2003; Marschner 2008).

Some microorganisms increase the solubility of insoluble tricalcium phosphate, dicalcium phosphate, hydroxyapatite, and rock phosphate. Various strains of *Azotobacter*, *Enterobacter*, *Salmonella*, *Rhodococcus*, *Rhizobium, Bacillus,* and *Pseudomonas* genera of bacteria and *Aspergillus, Penicillium, Saccharomyces, Fusarium, Arthrobotrys and Rhizoctonia* genera of fungi are among the phosphate solubilizing microorganisms (Alori et al., 2017). There is a significant microorganism population in soil and plant rhizosphere that can solubilize phosphate. Such microorganisms decrease the pH of the medium by secretion of organic acids and thus increase the solubility of phosphates. Fertilizers containing such microorganisms are also suggested for alkaline soils which contain insoluble phosphate (Bhargava et al., 1993, Mahdi et al., 2010).

Potassium and Bacteria

The main sources of potassium in soil are primary minerals such as potassium feldspar ($KAlSi_3O_8$), muscovite [$H_2KAl_3(SiO_4)_3$], and biotite [$(H,K)_2(Mg,Fe)_2Al_2(SiO_4)_3$]. Potassium is also found in the structure of illite, vermiculite, chlorite, and amorphous secondary clay minerals (Tisdale et al., 1993). A significant portion of potassium is found in the structure of primary soil minerals and potassium appears as these minerals disrupt in time.

There are many factors affecting the potassium uptake of plants (Rhem and Schimit 1997).

The main factors are:

- ✓ O_2 level- oxygen is required for root development and function which is effective on potassium uptake.
- ✓ humidity – potassium uptake is negatively affected by the drying of soil
- ✓ tillage - potassium uptake by plants increases in soils which are regularly tillage
- ✓ soil temperature -15.5-26 °C soil temperature is optimum for root activity and physiological processes. Below or higher temperature results in lower uptake.

The main metabolic effects of potassium in plants are that they activate several enzyme systems, increase durability against drought by enhanced root growth, decreases water loss and fading by maintaining cell turgor, take place in the opening and closing of stoma, inhibits energy loss by decreasing respiration, increase stoma activity and photosynthesis, transport of sugar and starch, increase starch content of grain, increase protein content of the plants, support the upright position of plants by increasing cell wall stability, increase the resistant of plants to pests and diseases. (Clarkson and Hanson 1980,

Walker et al., 1996, Mengel and Kirkby 2001, Britto and Kronzucker 2008, Marschner 2008, Maathuis and Sanders 2009). As a result, yield and quality increase in agricultural production.

Potassium fertilizers (bio-potash) contain *Bacillus* species. These microorganisms degrade potassium compounds and bring out the soluble K^+ ions. They significantly increase the yield when applied to soils with potassium deficiency (Fredee et al., 1989, Karaman et al., 2012).

Conclusion

Sustainable agriculture is becoming increasingly important to feed the world's growing population. Different farming systems and farming activities mainly determine the acquisition of agricultural products. However, unconscious agricultural practices result in the reduction of soil productivity and soil quality. In order to improve these properties and obtain the highest yield per unit area, maintaining soil productivity is a must. Therefore, soil quality parameters should be well-known and determined. For this purpose, natural and biological aids come into prominence. Biofertilizers are the foremost of such aids. Bacteria and mycorrhiza are the basis of biofertilizers. With the help of these microorganisms, the nutrient capacity of the soil, the physical, chemical, and biological parameters of the soil can be increased. In this context, scientific and technological research should be developed and supported in order to form new biofertilizer mixtures.

References

Adiloğlu, A. and S. Adiloğlu. 2021. Stella Rosa (Editor) Book title: *Boron: Advances in Research and Applications.* Chapter title: "Boron in Sustainable Soil Fertility" Nova Science Publishers, Inc.

Adiloğlu, S. 2020. Interaction of Some Heavy Metals with Copper Content in Dock Plant. *KSÜ Journal of Agriculture and Nature* 2 (4):1078-1084.

Adiloğlu, S. and M. T. Sağlam. 2015. Some Statistical Relationships Between Chrome and Nickel Contents and Some Physicochemical Properties of Tekirdağ Province Soils. *Journal of Tekirdag Agricultural Faculty* 12(2): 110-119.

Adiloğlu, A. and S. Adiloğlu. 2005. Fertilizer Consumption in Turkey Between 1960 and 2002. *Bulgarian Journal of Agricultural Sciences* 11(3): 317-321.

Arun, K. S. 2007. *Bio-fertilizers for sustainable agriculture. Mechanism of P-solubilization. Sixth edition,* Agribios publishers, Jodhpur, India (2007), pp. 196-197.

Adiloğlu, S., Eryilmaz Açıkgöz F., Irmak Yılmaz F., Solmaz Y., A. Adiloğlu. 2019. The Effects of Biochar and Raising Mycorrhiza Usage on Several Mineral Ingredient of Pak Choi. *Fresenius Environmental Bulletin.* 28(3): 2220-2225.

Akay A., Adiloğlu, S. 2020. Effects of Mycorrhizal Inoculation on the Zn Phytoremediation Potential of Buckwheat (*Fagopyrum esculentum*). *Fresenius Environmental Bulletin.* 29(08):6431-6437.

Adiloğlu, S., Eryilmaz Açıkgöz F., Bellitürk K., Gürgan M. Solmaz Y., and A. Adiloğlu. 2021. The effects of increasing amounts of vermicompost and a fixed amount of *Rhodobacter capsulatus* applications on macro and micro elements of plant and soil samples *Journal of Plant Nutrition.* 44(19): 2876-2884.

Alori, E. T., Glick, B. R. and O. O Babalola. 2017. Microbial Phosphorus Solubilization and Its Potential for Use in Sustainable Agriculture. *Frontiers in Microbiology* 8: 1-8.

Bhargava, B. S., Singh, H. P. and K. L. Chadha. 1993. Role of potassium in development of fruit quality. In: *Advances in Horticulture, 2 Fruit Crops: Part 2.* (Eds. K. L. Chadha and O. P. Pareek). Malhotra Publishing House, New Delhi, 947-960.

Britto, D. T. and H. J. Kronzucker. 2008. Cellular mechanisms of potassium transport in plants. *Plant Physiol.* 133:637-650.

Clarkson, D. T. and J. B. Hanson. 1980. The mineral nutrition of higher plants. *Annu. Rev. Plant Physiol.* 31: 239-298.

Eckert, B., Weber, O. B., Kirchhof, Halbritter, A., Stoffels, M. and A. Hartmann 2001. .*Azospirillum doebereinerae* sp. nov., anitrogen-fixing bacterium associated with the C4-grass Miscanthus *International Journal of Systematic and Evolutionary Microbiology* .51:17–26.

Fredee, A. L., Rao, I. M. and N. Terry 1989. Influence of phosphorus nutrition on growth and carbon partitioning in glycine max. *Plant Physiol.* 89: 225-230.

Gurgan, M., and S. Adiloğlu. 2021a. James C. Flores (Editor) Book title: *The Future of Phytoremediation.* Chapter title: "Phytoremediation and Purple Non Sulfur Bacteria" Nova Science Publishers, Inc.

Gurgan M. and S. Adiloğlu 2021b. Editor : Lynn Mcconnell, Book title: *Boron: Advances in Research and Applications.* Chapter title: "Remediation Of Boron Toxicity Using Bacteria" Nova Science Publishers, Inc.

Gürgan, M. and S. Adiloğlu 2020. Karoline Møller (Editor) Book title: *Brassica napus: Cultivation and Uses.* Chapter title: *Brassica napus*: Bioremediation. Nova Science Publishers, Inc. pp, 134.

Karaman, M. R., Aydın A., R., Brohi, A., Güneş, A., İnal, M., Kaplan, V., Katkat, A., Korkmaz. 2012. *Plant Nutrition.* Ankara : Dumat Ofset.

Koca, Y. K., Derin A. and S. Adiloğlu. 2019. Mapping of Some Micro Element Levels with Geostatistical Modeling in Uzunköprü District of Edirne Province. *Tekirdağ Journal of Agriculture Faculty* 16(3): 1-10.

Lowenfels J and W. Lewis 2010 *Teaming with Microbes : The Organic Gardener's Guide to the Soil Food Web*, Timber Press.

Maathuis, F. J. and D. Sanders. 2009. Mechanism of high-affinity potassium uptake in roots of Arabidopsis *Proc. Natl. Acad. Sci.* 91:9272-9276.

Marschner, H. 2008. *Mineral Nutrition of Higher Plants. Digital Print.* Academic Press., pp. 889.

Mengel, K. and E. A. Kirkby. 2001. *Principles of Plant Nutrition*. 5th Edition. Kluwer Academic Publishers. Dordrecht, The Netherlands.

Müftüoğlu, N. M., Türkmen, C. and Y. Çıkılı 2012. *Soil and Plant Productivity Analysis*. Sertifika No:11413, s. 216, Kriter Yayınevi, Fatih, İstanbul.

Pandey, P. K. and N. S. Sreelu 2006 Nitrogen-fixing anaerobic bacteria and their nitrogenase activity in North Mumbai coastal waters. *J. Indian Fish. Assoc.,* 33: 95-102.

Rehm, M. and M. Schmitt 1997. *Potassium for Crop Production. Nutrient Management.* University of Minnesota Extension.

Sessitsch, A. Weilharter, A., Martin H. Gerzabek, M. H., Kirchmann, H and E Kandeler. 2001. Microbial Population Structures in Soil Particle Size Fractions of a Long-Term Fertilizer Field Experiment. *Appl. Environ. Microbiol.* 67(9): 4215–4224.

Tok H. H., 1997. *Soil Biology*. T.Ü. Tekirdağ Agriculture Faculty, Class book. Tekirdağ.

Chapter 7

The Effects of Soil Management on Rhizosphere Dynamics

Aline Martineli Batista[1,2]
Antonio Marcos Miranda Silva[2]
Victor Lucas Vieira Prudêncio de Araújo[2]
and Thaís Nascimento Pessoa[2,*]
[1]Faculty of Agricultural Sciences, National University of the Littoral, Esperanza, Santa Fe, Argentina
[2]Department of Soil Science, University of São Paulo, Piracicaba, São Paulo, Brazil

Abstract

Rhizosphere is defined as the volume of soil that surrounds and influences plant roots. It is considered the most active portion of the soil and depends on three factors: plant genotype, soil (or substrate) properties, and microbial population. Processes occurring in the rhizosphere lead to changes in soil properties. For this reason, some research refers to them as "rhizospheric effects." Roots release molecules or other substances in a process called rhizodeposition. Among the many substances released by plant roots are root exudates and mucilage. Root exudates and other substances are rich in carbon and are closely related to the increase in microbial activity and selection of the microbial population in the rhizosphere. Mucilage plays an important role in hydraulic changes in this environment. Substances released by roots cause chemical changes, and together with biological changes, lead to many important biogeochemical processes in the rhizosphere. They are also responsible for soil physical changes in the soil of the rhizosphere.

* Corresponding Author's Email: thaisnpessoa@usp.br

In: Agricultural Soils
Editors: R. Nogueira de Sousa, T. Nascimento Pessoa and K. Ferreira Mendes
ISBN: 979-8-89113-000-5

Recent studies indicate that we can use the knowledge of changes in the rhizosphere to avoid stress conditions and improve plant growth and health. Understanding these changes for different cropping systems and in different environments will help in the selection and development of management practices that will increase plant health and sustainability in crop production, avoid stress situations, and help in the selection of crops or intercropping systems to improve soil quality.

Keywords: Soil-root interaction, microbial activity, chemical changes, biogeochemical processes, physical-hydric properties

Introduction

The rhizosphere is defined as the portion of the soil adjacent to a plant's root system and is influenced by physical (e.g., soil type, texture, moisture, and temperature), chemical (e.g., soil pH, electrical conductivity, fertilizer inputs, and nutrient status), and biological (e.g., plant succession, root system, and enzymes) factors. It is considered one of the most complex ecosystems on Earth due to its high variability (Jones et al. 2008; Hinsinger et al. 2009). Several authors try to delineate the rhizosphere by measuring it, but due to the relevant spatial and temporal scale resulting from the variability of the aforementioned factors, it can vary significantly between different plant species and phenological phases (Vetterlein et al. 2020).

The modulation of the rhizosphere depends on environmental conditions and results from the communication between plants and soil through rhizodeposition. Plant roots secrete metabolites into the rhizosphere that influence various processes in the soil. Apparently, the root exudation process is the main evidence of this communication because it alters the plant-soil interface in time and space and can attract beneficial microorganisms. These microorganisms play an important role in the biogeochemical process to improve plant growth and development and, in some cases, to avoid pathogenic attack.

Processes occurring in the rhizosphere typically stimulate weathering and influence soil formation. The rhizosphere provides an interface between biota and the geological environment, exerting physical pressure and creating a chemical environment in which biogeochemical reactions interact with minerals. Roots grow and expand in volume, pressing into the surrounding soil and into planes of rock weakness (e.g., joints or fractures). Root growth facilitates mechanical weathering of minerals, accelerates chemical weathering, and increases the specific surface area of minerals, promoting contact with microorganisms and organic compounds. Thus, the rhizosphere is essential for soil formation, including highly weathered soils such as Oxisols (Richter et al. 2007).

Understanding the interactions between plants and microorganisms in the rhizosphere is important because plant roots can select microorganisms from the wide range of chemical compounds released. The rhizosphere can modulate management decisions, tillage practices, development of agricultural products (e.g., fertilizers, biofertilizers, pesticides), and improvement of soil health (Huang et al. 2014). Plants can modify their growth environment and thus soil properties.

Rhizodeposition

The process of releasing carbon compounds from living roots is called rhizodeposition. This term was first described by Whipps and Lynch (1985) as the material lost from plant roots, including water-soluble exudates, secretions of insoluble materials, lysates, fine dead roots, and gases (Cheng and Gershenson 2007). Overall, rhizodeposition accounts for 17 – 20% of the total carbon fixed by photosynthesis, with the most recognized release mechanisms being the most recognized: (1) sloughing-off of root border cells, (2) secretion of mucilage by roots, (3) leakage of solutes from living cells, (4) senescence of the root epidermis, and (5) gas losses (Jones et al. 2009). Certainly, the amount and type of rhizodeposits are highly context specific, e.g., they vary with plant species (e.g., phenological and physiological status and root architecture), plant succession, soil properties (e.g., chemical, physical, and biological), atmospheric conditions (e.g., temperature, irradiance, and CO_2 content), and so on (Jones et al., 2009). The most common organic compounds released by plant roots are summarized in Table 1 according to Uren (2009).

The largest fraction of non-volatile rhizodeposits is reported to be root exudates (Meharg and Killham, 1988). Root exudates act as messengers that stimulate physical and biological interactions in the rhizosphere (Channarayappa and Biradar, 2018) and are closely associated with changes in soil properties in this environment (Chen et al. 2019b; Cruz et al. 2017; van Veelen et al. 2018). These root exudates modify soil properties, such as soil pH, to solubilize nutrients, chelate toxic compounds, and attract beneficial microorganisms (Vives-Peris et al. 2020). On the one hand, beneficial bacteria can fix nitrogen and mycorrhizal fungi can improve plant performance. On the other hand, pathogens can suppress plant growth (Hu et al. 2018).

Table 1. List of organic compounds released by plant roots with examples of chemical compounds provided by Uren (2009)

Functional group	Chemical compounds
Sugars and polysaccharides	Fructose, glucose, sucrose, xylose, galactose, maltose, mannose, oligosaccharides, etc.
Amino acids	Arginine, asparagine, cysteine, glutamine, leucine, lysine, methionine, phenylalanine, proline, serine, threonine, tryptophan, tyrosine, valine, etc.
Organic acids	Ascorbic, benzoic, butyric, fumaric, glutaric, glycolic, glyoxilic, lactic, malic, malonic, oxalacetic, oxalic, etc
Fatty acids	Linoleic, linolenic, oleic, palmitic, stearic
Sterols	Campesterol, cholesterol, sitosterol, stigmasterol
Growth factors	*p*-amino benzoic acid, biotin, choline, niacin, pantothenic, vitamins B1 (thiamine), B2 (riboflavin), and B6 (pyridoxine)
Enzymes	Phosphatases, amylase, invertase, peroxidase, protease, etc
Flavonones and nucleotides	Adenine, flavonone, guanine, uridine or cytidine
Miscellaneous	Auxins, hydrocyanic acid, glucosides, hydroxamic acids, ethanol, glycinebetaine, inositol and myo-inositol like compounds, etc.

Table 2. Plant mucilage components and the functions and proportion of each component in the rhizosphere

Mucilage components	Functions
Polysaccharides (78.4%)	• Energy source for microorganisms • Water absorption • Exchange of cations • Adherence to solid surfaces (e.g., soil particles and • nematodes)
Proteins (7.3%)	• Disassembly of the mucilage polysaccharide for microbial use • Response to abiotic and biotic stresses (e.g., heat stress and plant pathogens) • Assistance in nutrient uptake (e.g., phosphorus and iron)
Minerals (5.6%)	• Exchange of cations • Improvement of soil liquid-phase connectivity and water content
Lipids (3.1%)	• Alternation of soil hydraulic properties • Desorbing adsorbed phosphorus on soil particles • Easing the penetration of roots into the soil

Source: Nazari (2021).

Mucilage, a polymeric gel released by roots, is also responsible for many changes in the soil of the rhizosphere (Zarebanadkouki et al. 2016). In a generalized way, Table 2 shows the main functions of each of the main components of the mucilage.

Rhizodeposition can function as inter-root and root-microbial signal exchange, therefore, changing the rhizosphere microbial community (Jones et al. 2004; Paterson et al. 2007). In this regard, we refer the reader to the review by Paterson (2003) and Singh et al. (2004) since these authors delve into the regulation of microorganisms in the rhizosphere by rhizodeposition. Furthermore, we strongly recommend reading the second edition of the book *The Rhizosphere: Biochemistry and Organic substances at the soil-plant interface*, edited by Roberto Pinton, Zeno Varani, and Paolo Nannipieri, which, throughout its 14 chapters, addresses the main issues related to rhizodeposition and its interactions with the soil and plant systems. Finally, it is also inviting to go through the insights provided by Gregory et al. (2022) on recent methodological methods for understanding the rhizosphere and rhizodeposition.

Soil Organisms in the Rhizosphere

Soil organisms play a key role in maintaining the balance between inputs and outputs in the soil-plant-atmosphere system, providing ecosystem services, and therefore ensuring soil health. Although there are a myriad of soil organism classifications (e.g., based on the body size, metabolism, function, and so on), this chapter will highlight the microbes (mycorrhizal fungi, bacteria, archaea, nematodes, and virus) associated with the plant rhizosphere (Figure 1).

Biochemical processes carried out by microbial activity are catalyzed by enzymes, which are specific proteins that cause an acceleration in the rate of chemical reactions. Enzyme activity in the rhizosphere can be used as an index of microbial functional diversity, as long as it reflects vital microbial processes. Some enzymes are present in almost every microbial cell, such as phosphofructokinase-1, one of the 10 regulatory enzymes of glycolysis, which catalyzes the conversion of fructose-6-phosphate into fructose-1,6-bisphosphate (Starr et al. 2018).

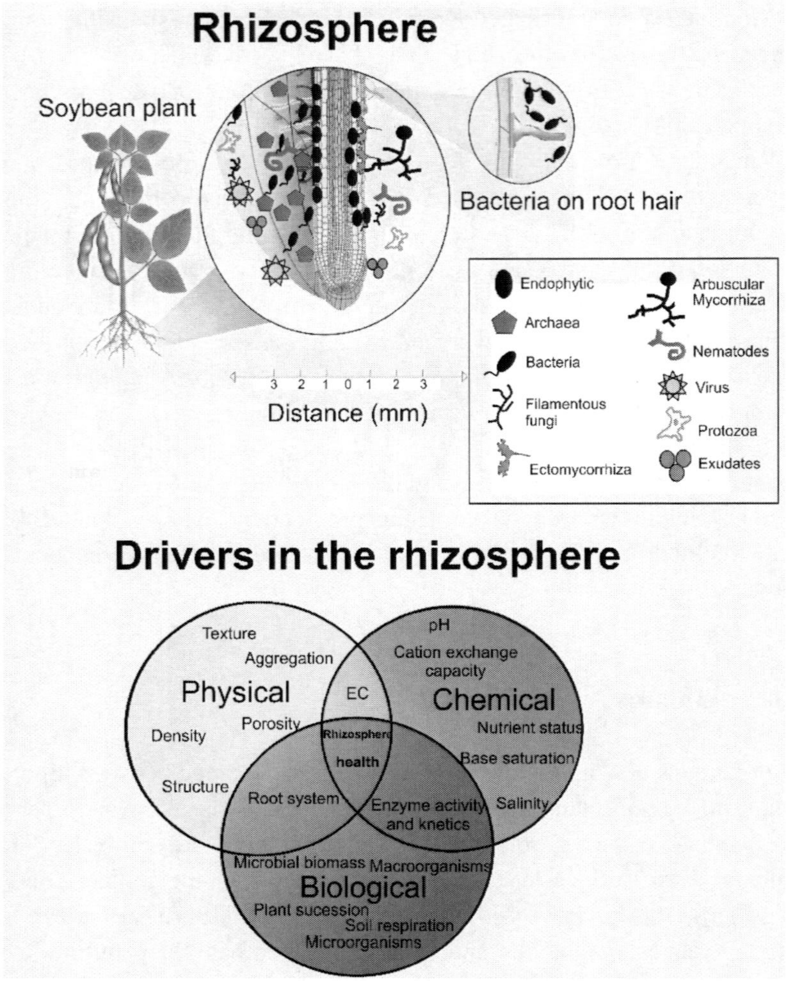

Source: original draft granted by Yasmin F. Rodrigues.

Figure 1. Schematic representation of the rhizosphere showing the microbial composition, bacteria colonizing the root hair and the main physical, chemical and biological in the rhizosphere. EC: electrical conductivity.

Plants and microorganisms are the main sources of enzymes in the soil, either by direct synthesis and secretion or as signaling agents that induce enzyme synthesis by other microorganisms or plants. Enzymes such as alkaline phosphatase and ribonucleases, for example, are widely known and secreted by *Bacillus*. On the other hand, pyrophosphatases and acid phosphatases are widely known to exist on the outer surface of the cell wall of

Saccharomyces. In addition, plants are capable of releasing acid phosphatases, amidases and ureases, for example (Burns and Dick 2002). Scientific evidence has shown the role of soil enzymes in reflecting the soil health status and related to yield increases, representing an advance in the understanding of the so-called black box of microbial contributions (Mendes et al. 2021; Silva et al. 2023).

Before discussing into the subject of microbial activity in the rhizosphere, it is important to define some ecological terms, such as microbial richness, uniformity, composition, diversity, and functional microbial diversity. Microbial richness is understood as the number of different species in a community. Whilst microbial uniformity is the relative contribution that individuals of all species make to the total number of organisms present. On the other hand, microbial composition is the type and relative contribution of each species present. Finally, functional microbial diversity refers to both substrate utilization rates and the presence/absence of specific substrate utilization. A decrease in diversity can reduce the microbial functionality of the rhizosphere if key species are affected. Furthermore, in the rhizosphere, microbial groups can perform multiple similar functions and therefore, which is referred to as functional redundancy (Escalas et al. 2019).

In natural ecosystems, the microbial diversity is higher in non-rhizospheric soils (bulk soil – no direct root interference) than in rhizospheric soils. On the other hand, the total abundance of beneficial microbes, as well as the microbial biomass and network complexity, is higher in the rhizosphere than in the bulk soil (Sokol et al. 2022). This is due to the recruitment made by the plant against the diversity of the bulk soil, ensuring the so-called symbioses such as mycorrhizae, nitrogen-fixing bacteria, phosphate-solubilizing bacteria, etc. According to Mendes et al. (2013), the rhizosphere is colonized by different microbial groups called the good (beneficial microbes), the bad (plant pathogens), and the ugly (human pathogens), resulting in the rhizosphere microbiome (Figure 2). However, what generally distinguishes these categories is the abundance found in the rhizosphere; likewise, the difference between poison and medicine is in the dose (Hartmann et al. 2008; Mendes et al. 2013).

Mycorrhiza (myco = fungus and rhiza = root) is considered to be part of the good group in the rhizosphere and was first mentioned in 1885 by the botanist Albert Bernhard Frank who described it as the symbiotic association between fungi and plant roots, dating back more than 400 million years. Seven types of mycorrhizas are recognized, including arbuscular mycorrhiza, ectomycorrhiza, orchidoid, ericoid, arbutoid, ectendomycorrhiza, and monotropoid. These types differ in function and anatomical structure, for example, the orchidoid category is associated with plants without chlorophyll (achlorophyllous plants), in which fungi act by supplying carbon to the plants while the plants provide nutrients to them (Smith and Read 2010).

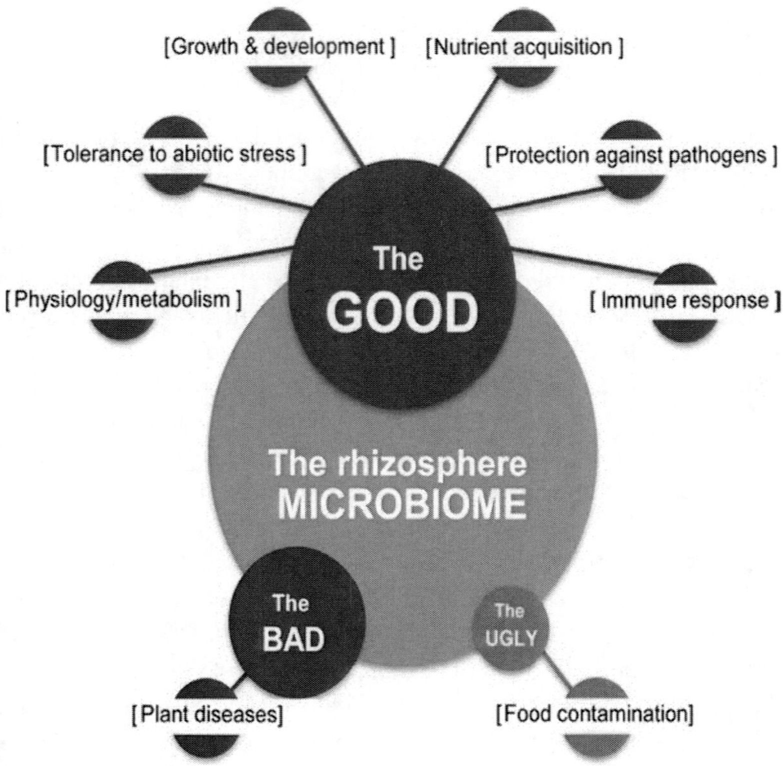

Source: Adapted from Mendes et al. (2013).

Figure 2. Overview of the rhizosphere microbiome, showing microbial groups that promote plant growth and nutrition (the good), cause plant diseases (the bad), or human disease/contamination (the ugly).

On the other hand, arbuscular mycorrhizal fungi (AMF) and ectomycorrhizal (ECM) are the most abundant types that directly participate in the plant nutrition, while plants supply them with the photoassimilates. AMF, in contrast to ECM, are obligate biotrophs because they cannot complete their life cycle without a host plant. In summary, mycorrhizae are know to increase the water and nutrients uptake by plants, especially P, N, Zn, Cu, and K. In addition, AMF represents a strategy to increase carbon storage in the soil due to their vast hyphal extension. Futhermore, AMF synthesize a glycoprotein called glomalin which, in addition to creating a form of carbon storage, acts as a cement for soil particles and, together with the extraradical mycelium, provides better soil structuring, especially in the stabilization of macroaggregates (>250μm) (Siqueira et al. 2010).

AMF is considered a traveler due to its high growth (reaching km) underground to provide nutrients that plant roots could not capture on their own. Therefore, these particularities make AMF a biological alternative for use in agriculture, especially in systems that seek greater agricultural sustainability, where the reduction of synthetic inputs (fertilizers and pesticides) is needed.

Another component of the beneficial group in the rhizosphere is the bacteria that are able to protect or promote plant growth through biochemical mechanisms. The group of bacteria intimately associated with the plant rhizosphere are called plant growth promoting rhizobacteria (PGPR) (Mohanty et al. 2021). In analogy to fungal root symbionts, Lorenz Hiltner, who was the first scientist to describe the term rhizosphere, named the bacterial community of the rhizosphere as bacteriorhiza and argued that these beneficial bacteria are not only attracted by the root exudates, but are also "uninvited guests" that adapt to the specific root exudates (Hartmann et al. 2008).

Therefore, the rhizosphere soil must be faced as our human body, where it is necessary to be aware that the number of bacterial cells may or may not exceed the number of human and/or plant cells (Abbott 2016; Sender et al. 2016). Our living bacteria are responsible for almost all the biochemical processes that allow us to breathe, walk, speak, and make our planet habitable. Likewise, these bacteria make it possible to transform the leaves that fall from the trees into available nutrients for the plant and other microbes (Backer et al. 2018). In addition, bacteria allow communication either between plants, alerting them to possible invaders, or between rhizosphere bacteria, controlling their population by using small signaling molecules (a process called quorum sensing) or disrupting communication via degradation of these molecules (a process called quorum quenching) (Rehman and Leiknes 2018). The benefits provided by bacteria result from direct and indirect mechanisms of action, some of which are summarized in Table 3.

Table 3. Direct and indirect mechanisms of action provided by bacteria to promote plant growth or protection

Direct mechanisms*		
Mechanism of action	PGPR	Reference
Biological nitrogen fixation (BNF)	*Bradyrhizobium* spp. *Azospirillum brasilense*, *Azospirillum lipoferum*	Hungria et al. (2010) Barbosa et al. (2021)
Phosphate solubilization	*Bacillus* sp. *Rhizobium* sp. *Herbaspirillum* sp.	Estrada-Bonilla et al. (2021) Santos-Torres et al. (2021) Romero-Perdomo et al. (2021)
Auxin production	*Klebsiella planticola*	Blinkov et al. (2014)
Cytokinin production	*Frankia* sp.	Nouioui et al. (2019)
Gibberellin production	*Azotobacter chroococcum*	Zhang et al. (2019)
ACC- deaminase	*Bacillus cereus*, *Bacillus wiedmanni*, *Bulkhorderia cepacian*	Araújo et al. (2020) Maxton et al. (2018)
Heavy metals uptake	*Pseudomonas* spp. *Arthrobacter* spp.	Mishra et al. (2017)
Volatile organic compounds (VOCs)	*Bacillus subtilis*	Tahir et al. (2017)
Indirect mechanisms		
Mechanism of action	PGPR	Reference
Siderophores production	*Azotobacter chroococcum*	Zhang et al. (2019)
Cell wall degrading enzymes	*Paenibacillus* sp. *Streptomyces* sp.	Singh et al. (1999)
Induced systemic resistance (ISR)	*Pseudomonas fluorescens*	Verhagen et al. (2004)
Antibiotics production	*Pseudomonas* sp.	Mazurier et al. (2009)
Oxidative stress management	*Bacillus* spp.	Moreno-Galván et al. (2020)

*A detailed understanding of direct and indirect mechanisms can be seen in the revision by Glick (2012).

Before shedding light on the next good group of microorganisms that inhabit the rhizosphere, it is crucial to note that in 1977 the concept of life in evolutionary biology was divided into three domains, e.g., Archaea, Bacteria, and Eucarya. Previously, it had been divided into prokaryotes and eukaryotes (Woese and Fox 1977; Cox et al. 2008). Although archaea and bacteria possess a unique cell without a nucleus (prokaryotic characteristics), they have a different cell wall composition with distinctive metabolic pathways and enzymes (Moissl-Eichinger et al. 2018; Bruns 2021). Furthermore, it is worth noting that since the discovery of the structure of DNA by James Watson and Francis Crick in 1953, the microbial community of rhizosphere has been continuously revealed until now (Watson and Crick 1953; Suzuki 2020).

Despite the scarce information on archaeal communities compared to bacteria, it is known that they represent a key living microbial group in the plant rhizosphere, providing nutrients to the plant and promoting growth and resistance, especially in extreme environments (Moissl-Eichinger et al. 2018; Jung et al. 2020). The most recognized archaeal phyla are the Crenarchaeota, Euryarchaeota, Thaumarchaeota, Aigarchaeota, and Korarchaeota (Eme et al. 2017). For example, in the rice rhizosphere, ammonia-oxidizing archaea, belonging to the phylum Thaumarchaeota, were found to be dominant due to the root exudate composition (Chen et al. 2008). However, it is important to consider that the archaeal community varies depending on the associated plant species, soil edaphoclimatic conditions and phenological stage (Jung et al. 2020; Zhang et al. 2020). Research on archaeome and plant–archaea interactions is still ongoing, and the main obstacle is based on methodological limitations (Moissl-Eichinger et al. 2018).

Belonging to the domain Eucarya, as well as the arbuscular mycorrhizal fungi mentioned above, nematodes are characterized as a good group forming the rhizosphere community (Wilschut and Geisen 2021). Although nematodes are mostly recognized as plant pathogens, only about 7% (ca. 4100 species) can cause plant diseases, but this represents serious damage and yield losses worldwide, reaching 10-25% in many crops (Decraemer and Hunt 2006; Mathesius and Costa 2021). The ten most important plant parasitic nematodes, such as those belonging to the rhizosphere nematode group, are summarized in Table 4 according to Jones et al. (2013).

According to Gebremikael et al. (2016), nematodes are able to enhance plant growth and nutrient uptake by plants under eutrophic conditions, as long as their abundance is low. This improvement is driven by direct and indirect effects, through interactions with other microorganisms in the rhizosphere, such as bacteria, archaea, and arbuscular mycorrhizal fungi. So far, the nematobiome assessment in search of keystone species that improve plant health has been neglected, mainly due to the lack of advanced methodological techniques, which is the main bottleneck in nematode studies (Wilschut and Geisen 2021).

For instance, it is important to enhance the interaction between nematodes and the microbial community. For example, Barros et al. (2022) observed in soybean that microbial diversity was able to reduce the population density of plant-parasitic nematodes (*Meloidogyne javanica*), and thus improve plant performance. Likewise, Zhou et al. (2019) observed that microbial diversity was higher in the absence of *Meloidogyne* spp., than in soils with the presence of these nematodes, in addition to the microbial communities exclusive to these soils. In tobacco plants, Huang et al. (2020) showed that the occurrence of the disease caused by *Meloidogyne* spp. was associated with changes in soil microbial communities, mainly due to interactions among *Pseudomonas*, *Bryobacter*, *Variibacter*, *Coniochaeta*, and *Metarhizium*.

Table 4. Top 10 plant-parasitic nematodes according to Jones et al. (2013), showing the main damage, nematodes examples, and plant host

Damage caused	Nematodes	Host*
Root-knot nematodes	*Meloidogyne* spp.	Parasitize almost every species of vascular plant (e.g., *Solanum lycopersicum*, *Oryza sativa*)
Cyst nematodes	*Heterodera* spp. *Globodera* spp.	*Glycine max* *Solanum* sp.
Root lesion nematodes	*Pratylenchus* spp.	*Triticum aestivum* *Saccharum* spp. *Coffea* sp. *Zea mays*
The burrowing nematode	*Radopholus similis*	*Citrus* sp. *Capsicum* sp. *Musa* sp.
The stem and bulb nematode	*Ditylenchus dipsaci*	*Oryza sativa* *Solanum* sp. *Arachis hypogaea* *Fragaria* spp.
The pine wilt nematode	*Bursaphelenchus xylophilus*	Parasitize especially trees (e.g., *Tectona grandis*)
The reniform nematode	*Rotylenchulus reniformis*	Parasitize vegetables, fruits, ornamentals and fibre crops, as well as weeds (e.g., *Vigna unguiculata*, *Gossypium* sp., *Ricinus communis*)
The only virus vector nematode to make the list	*Xiphinema index*	*Vitis* sp.
The false root-knot nematode	*Nacobbus aberrans*	*Solanum* sp. *Capsicum frutescens*
Causes white tip	*Aphelenchoides besseyi*	*Oryza sativa*

*Just a few host examples.

Finally, yet importantly, viruses are the next rhizosphere group to be mentioned, which are also often related to plant diseases such as nematodes. By definition, a virus is a microbe consisting of a segment of nucleic acid surrounded by a protein coat that lives in infected cells and produces virions to spread itsgenes (Forterre and Prangishvili 2009). Bi et al. (2021), investigating the virome of eight rhizosphere and bulk soils in an agroecossystem in southwestern China, conclude that the composition of the viral community differed significantly between bulk and rhizosphere soils, with soil pH being the main environmental driver of the viral community structure. In addition, these authors highlight that the order *Caudovirales* was the predominant viral type in agricultural soils, with *Siphoviridae* being the most abundant family. Some viruses are capable of infecting bacteria, and are therefore known as bacteriophages, which play a key role in plant health, especially due to the regulation of the host-associated bacterial communities (Pratama et al. 2020). As mentioned above, the rhizosphere microbiome should be thought of as the human body. Pratama et al. (2020) arrived at this last statement by drawing a parallel between the human gut microbiome and the plant rhizosphere, both environments considered to be the home of the microbes and influenced by physical, chemical, and biological characteristics. Furthermore, the authors compared the bacteriome and virome of the gut and rhizosphere exposed to disturbance, explaining that the bacterial communities are drastically affected by perturbations, while the virome seems to be resistant to or less affected, and this is due to phage persistence, the activation of prophages, or horizontal gene transfer.

Therefore, efforts to understand the rhizosphere community must focus on holistic approaches that include the virome, nematobiome, bacteriome, archeome, and mycorrhiza. In addition, it is worth mentioning that here we do not discuss here other meso- and macroorganismsliving in the rhizosphere, that are crucial for the maintaing soil and plant health. We strongly recommend reading the reviews by Oburger and Schmidt (2016) and York et al. (2016) for a very detailed reading on new methods to elucidate rizosphereprocesses and a holistic approach.

Chemical Properties in the Rhizosphere

In general, nutrient availability in the rhizosphere is controlled by several mechanisms, including root morphological responses, external inputs, organic matter, and exudation of organic compounds, factors that can be modulated by the microbial community of the rhizosphere (Mortimer et al. 2018; Schmidt et al. 2022). In summary, the rhizosphere is considered a dynamic environment where biochemical processes driven by microbial activity are extremely intense.

Roots promote changes in the chemical composition of the rhizosphere, such as changes in ion concentrations, pH changes in relation to the nutrient source, exudation of low molecular weight exudates (e.g., organic acids and amino acids), excretion of enzymes (e.g., phosphatase), and changes in redox potential (Gregory and Hinsinger 1999).

Soil pH is an important chemical property that affects nutrient solubility and availability to plants. Most micronutrients are highly available at low pH (< 5.5), except for molybdenum (Mo). The increase of pH ~7.0, may help in the availability of N and P, due to the higher activity of bacteria for the turnover of N, and the neutralization of positive charges associated with Al, Mn and Fe, which contribute to the adsorption of P. Soil pH in the rhizosphere are associated with changes in ions absorption and release of H+ or OH- (HCO3-) for plant roots (Fageria and Stone 2006).

A field experiment was conducted with different plants (i.e., millet, sorghum, cowpea, and groundnut) in sandy soils with P deficiency where the main aspects evaluated were the changes in pH and nutrient availability in relation to root distance and the application of a mineral fertilizer source. The pH increased ~ 0.7 and two units between bulk soil and millet rhizoplane, with a similar gradient in cowpea. In sorghum, pH changes were lower by ~ 0.3 units. These changes in pH increased the availability of nutrients close to the roots, mainly P. The millet rhizosphere also had a higher content of exchangeable calcium (Ca) and magnesium (Mg). The results showed that increasing the pH close to the roots might be important for the resilience of crops when the soil presents high acidity and exchangeable aluminum (Bagayoko et al. 2000). Song et al. (2007) showed the positive effect of intercropping on microbiological and chemical properties in the rhizosphere. They studied crop yield (wheat, maize, and faba bean) and intercropping (wheat/faba bean, wheat/maize, and maize/faba bean). Intercropping increased microbial biomass C, N, and P availability, but a reduction in microbial biomass N rhizosphere of wheat was detected.

Nitrogen (N) forms can affect the uptake of cations or anions and promote changes in the pH of the rhizosphere. N is available to plants such as NH_4^+ (ammonium) and NO_3^- (nitrate) which can account for ~ 80% of the uptake of ions. NO_3^- is the main form of N available to plants in most parts of well aerated tillage soils. When NO_3^- has a higher concentration in the soil solution, its absorption releases OH^-, and the pH increases. However, under conditions of higher NH_4^+ concentration, H^+ is released, and the pH decreases. Typically, NO_3^- predominates in oxidized soils, and in reduced soil conditions, NH_4^+ is the dominant ion in reduced soil conditions (Fageria and Stone 2006; Neumann and Römheld 2012).

Plants can adapt their nutrient acquisition strategies according to the N availability in the soil. Hao et al. (2022) cultivated maize at three fertilization rates to understand the changes in the exudation of the bacterial community in the rhizosphere. Fertilization reduced the activity of the enzymes b-1,4-glucosidase and b-N-acetylglucosaminidase by 48% and 39%, respectively, compared to zero fertilization rates. The decreased activity of hydrolases and their extension into the rhizosphere as a consequence of N fertilization were correlated with the relative abundance of oligotrophic (e.g., *Acidobacteria*) and copiotrophic (e.g., *Pseudomonadota* and *Patescibacteria*) bacteria, which increased with the addition of N. Thus, maize has adapted the composition of root exudates with the aim of promoting a higher bacteria/rhizosphere interaction and modulating the activity of hydrolases.

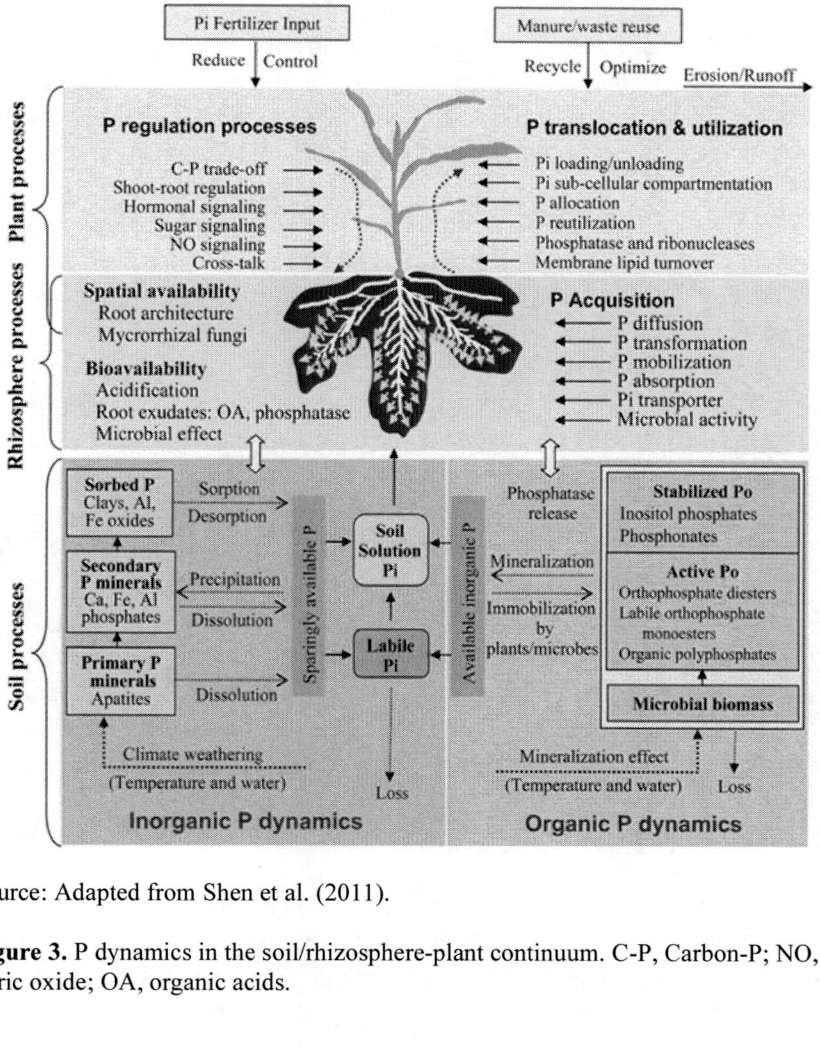

Source: Adapted from Shen et al. (2011).

Figure 3. P dynamics in the soil/rhizosphere-plant continuum. C-P, Carbon-P; NO, nitric oxide; OA, organic acids.

Phosphorus (P) usually has low availability and high adsorption in soil, especially in highly weathered soils. Adequate supply of P in the root zone can increase the efficiency of plant roots to mobilize and acquire P from the rhizosphere due to morphological and physiological adaptations in roots. P in soil occurs in different chemical forms, such as inorganic P (P_i), and organic P (P_o). P_i represents ~35 to 70% of the total P. The P from primary minerals (e.g., apatite, strengite, and variscite) has a slow release of P to attend to crop demand. However, the P from secondary minerals (e.g., calcium - Ca, iron - Fe, and aluminum - Al phosphates) have dissolution rates that vary with pH and mineral particle size. Increasing soil pH increases the solubility of Fe and Al phosphates but decreases the solubility of Ca phosphate (Figure 3). Thus, the availability of P to plants can be controlled, mainly for the spatial availability/acquisition of P, which depends on the association between root architecture and mycorrhizas, and the rhizosphere, which controls chemical and biological processes that affect the bioavailability/acquisition of P (Shen et al. 2011).

Liu et al. (2022a) studied the influence of different sources of mineral phosphate fertilizer rates (0, 60, 120 and 300 kg P_2O_5) on soil chemical properties and the bacterial community of the maize rhizosphere. The increase of fertilizer rates decreased the organic C concentrations, and fractions of Pi and Po were increased. The relative abundance of beneficial bacteria also increased with reduced fertilizer rates. The results showed that the lower rates of phosphate fertilizers may promote the beneficial bacteria and the diversity of these communities.

Liu et al. (2022b) used meta-analysis to investigate nutrient changes in the rhizosphere. From the 123 studies reviewed, the main findings were: (*i*) the increase in N cycling and enzyme activity in the rhizosphere increased available N by 10% compared to the bulk soil; (*ii*) plants and microorganisms competed for P, increased phosphatase activity, and decreased the available P by 12%; (*iii*) taproots concentrated more organic C (+17%) compared to fibrous roots, indicating a higher composition of fungi and bacteria in the rhizosphere of taproots; and (*iv*) the maximum accumulation of nutrients was higher in the rhizosphere when the soil was poorly fertile.

Carbon dioxide (CO_2) can be released from a root system and soil in several ways, the most important being root respiration, microbial respiration from root-derived materials, and microbial respiration using soil C. Such relationships are complex, making studies of root respiration and root exudation difficult. The zone around the rhizosphere can also have high rates of O_2 consumption due to root and microbial respiration. Respiration can interfere with the dynamics of O_2 concentration. In the soil, O_2 diffusion depends on the water content in the soil and on the reduction of water content in the soil pores around the rhizosphere, as plant evapotranspiration can cause an increase in O_2 diffusion rates. Thus, the total O_2 availability in the soil atmosphere of the rhizosphere depends on the soil water content in the soil and the rates of O_2 exchange with the soil and atmosphere (Cardon and Whitbeck 2007).

The main changes that occur in the rhizosphere are related to the dynamics of C between the rhizosphere and the soil. The flux of C and N in the rhizosphere varies in time and space, depending on the plant species and the environment around the root. All these factors influence the type of rhizodeposits present in the soil (e.g., mucilage, exudates) because it is complex, and the modeling of rhizodeposition and C sequestration has become a limiting issue.

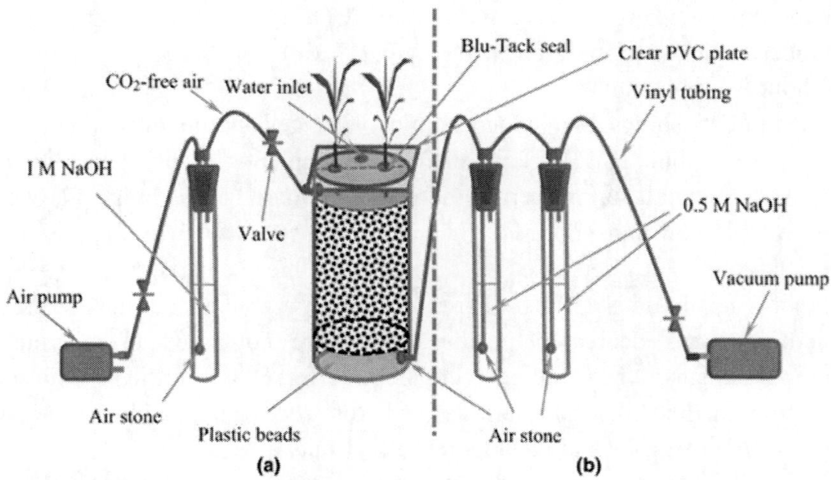

Source: Adapted from Wang et al. (2016).

Figure 4. Trapping system of CO_2 released from belowground; (a) leak checking, (a, b) CO_2 trapping. The CO_2 capture system has an inlet at the top of each column where CO_2 free air can be introduced after being pumped through 1 M NaOH via an air stone. At the base of the column, a vacuum line was installed at the outlet passing through two CO_2 collectors containing 150 ml of 0.5 M NaOH.

Wang et al. (2016) investigated the effect of rhizosphere properties on the rhizosphere priming effect (RPE). The authors evaluated four species of C3 plants (chickpea, field pea, wheat, and white lupine) with differences in soil acidification and root exudation grown in a soil cultivated with C4 plants. CO_2

released from the soil was captured using a newly developed NaOH capture system (Figure 4). The main results showed that: (*i*) white lupine and wheat had higher positive RPEs, due to their greater ability to release root exudates; (*ii*) negative RPEs were produced by chickpea, related to its high ability to acidify the rhizosphere soil. The main factor influencing the magnitude and direction of RPE was rhizosphere acidification, with a minor role assigned to biomass and root length.

Guo et al. (2022) investigated the effects of elevated CO_2 and co-elevated temperature on P availability and changes in rhizosphere microbial functions of soybean and rice grown on Mollisols with different P applications. A controlled environment was used to achieve a CO_2 concentration of 700 ppm and a temperature of ~ +2°C above ambient. Increases in P uptake were found in soybean in treatments without (23%) and with (28%) P application, and in rice by 34% (without P application) and 13% (with P application). It is likely that P mineralization occurs in soybean soils and P immobilization in rice soils under climate change. The co-elevation of CO_2 and temperature favored the number of P functional genes (e.g., *phoD*, *phoC*, *pstS* and *phnX*) in soils without P application.

Figure 5 shows the biogeochemical changes in the rhizosphere with respect to the bulk soil (left) and the nutrient depletion/accumulation process (right). In general, soil properties increased compared to the bulk soil, except for soil pH and phosphorus (P), which decreased by 0.16 units and 10%, respectively. The decrease in P indicates the consumption of P by microorganisms in the rhizosphere. The relative abundance of K strategist microorganisms decreased in the rhizosphere compared to r strategist microorganisms, which are very competitive for C and nutrients. Nutrients available in the rhizosphere are derived from the decomposition of organic matter, root exudates and metabolites, etc. (Liu et al. 2022b).

Finally, but importantly, the processes occurring in the rhizosphere also modify the redox potential and affect the forms of Mn (Mn^{2+} and Mn^{4+}), Fe (Fe^{2+} and Fe^{3+}), and Cu (Cu^{+} and Cu^{2+}), mainly Mn and Fe (Fageria and Stone 2006). Bravin et al. (2012) investigated root-induced changes in soil pH and dissolved organic matter (DOM), which modified the dynamics of Cu in durum wheat. The root-induced changes influenced the DOM, and, consequently, a decrease in Cu^{2+} concentration with an increase in soil Cu labile.

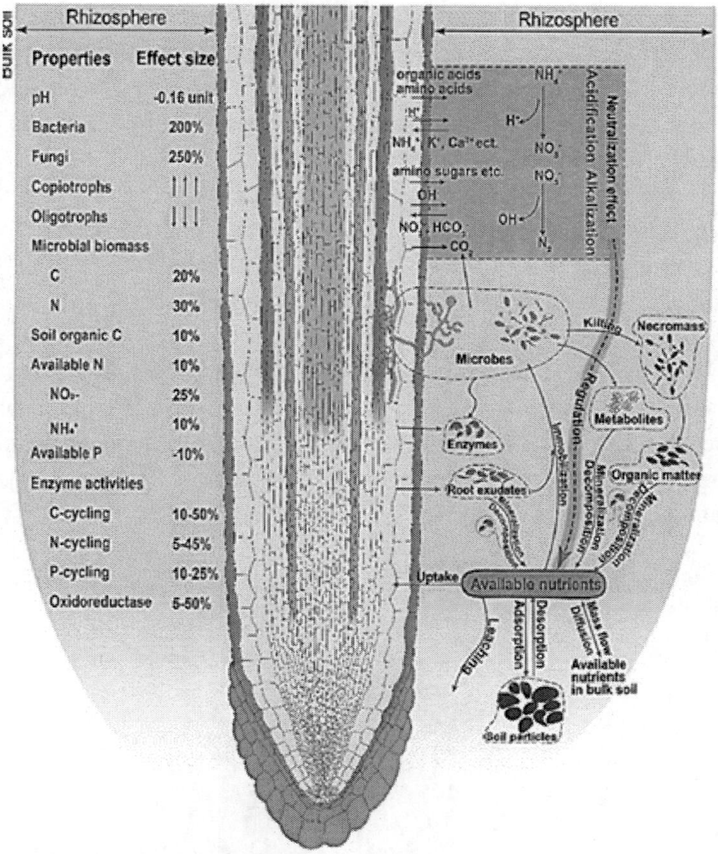

Source: Adapted from Liu et al. (2022b).

Figure 5. Changes of soil biochemical properties in the rhizosphere compared to bulk soil (left), and nutrient depletion and accumulation processes (right).

Soil Physical and Hydraulic Changes in the Rhizosphere

In the rhizosphere, soil aggregates are associated with root hairs (Figure 6) (Fortuna 2012). According to this author, extracellular polysaccharides, secreted mucilage and filaments of fungal hyphae act as binding agents in this environment. Thus, roots significantly influence the structure of the surrounding soil through the formation and stabilization of aggregates (Liu et al. 2019).

There is a strong relationship between the stability of soil aggregates and the microbial community inhabiting the soil, which is more abundant in the rhizosphere (Merino-Martín et al. 2021). As discussed above, roots can regulate the microbial community in their environment through the exudation of a variety of compounds (Walker et al. 2003; Kumar et al. 2017). The production of extracellular enzymes by microorganisms is largely responsible for the microbial decomposition of organic matter, and the activity of these enzymes is triggered by the presence of plants (Kumar et al. 2017). The association of plants with microorganisms favors the highest aggregate stability index in the rhizosphere in response to increased enzymatic activity (Hosseini et al. 2015; Demenois et al. 2018; Kuzyakov and Razavi 2019). When associated with mycorrhizal fungi that form mycorrhizal hyphae, an increase in soil aggregation is observed due to the release of glomalin, a glycoprotein involved in soil aggregate stabilization (Wright et al. 1996; Wright and Upadhyaya 1998; Demenois et al. 2018).

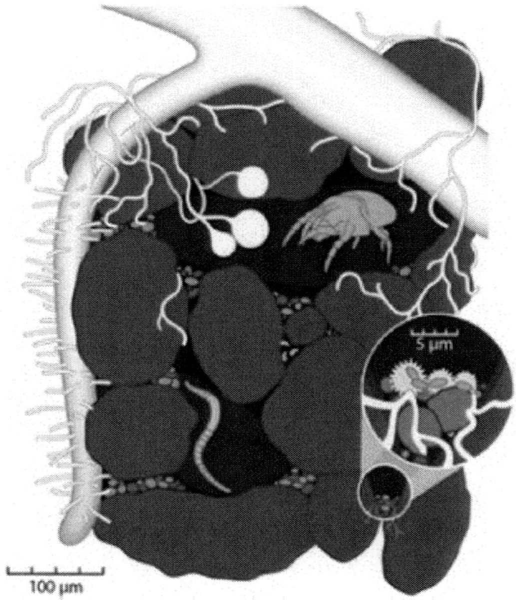

Source: Adapted from Fortuna (2012).

Figure 6. Soil aggregate formed naturally from mineral particles, organic matter, root hairs, microorganisms and their resulting secretions and pores.

Root-released mucilage is largely associated with improvements in soil aggregation and is linked to wetting and drying cycles and mycorrhizal hyphae formation, acting as a cohesive force in the rhizosphere (Cruz et al. 2017; Chen et al. 2019b; van Veelen et al. 2018). Because plant genotype quantitatively and qualitatively affects the release of the root exudates, which in turn affects the quality and composition of the rhizosphere microbiota, plants of different genotypes may have different effects on soil aggregation (Demenois et al. 2018; Graham e Strauss 2021).

On the one hand, it is widely accepted that the rhizosphere soil has greater aggregate stability than the soil without root influence due to the processes taking place in the rhizosphere. On the other hand, recent studies have shown that certain root exudates can lead to rhizosphere soil degradation, possibly due to the high concentrations of organic acids in certain soils (Naveed et al. 2017; Batista et al. 2020). These organic acids can destroy some cementing agents, such as iron, which is a very common cementing agent in highly weathered soils in tropical regions, leading to a reduction in aggregate stability. Thus, the substances released by the roots, the age of each plant, and biotic and abiotic stresses can flocculate or disperse the soil (Naveed et al. 2017).

During soil aggregation, the formation, stabilization, and disintegration of soil aggregates, and consequently their associated pore spaces, can occur (Lehmann et al. 2017). Few studies have assessed the pore space of rhizosphere soils, mainly due to the limitations of techniques to access porosity in situ without disturbing the soil (Helliwell et al. 2017). To directly access this property of the rhizosphere soil, it is necessary to use X-ray computed tomography. This makes direct determination expensive and exclusive to a few researchers.

The spatial and temporal arrangement of soil porosity is a key factor in soil functioning (Hinsinger et al. 2009). The distribution and size of pores are determined by soil structure and texture (Fortuna 2012). In the rhizosphere, root exudates influence root growth pressure and soil porosity (Oleghe et al. 2017). Whalley et al. (2005) showed that the soil structure of the rhizosphere differs from the soil structure without the influence of roots. Therefore, when soil aggregation and structure are altered in the rhizosphere, pore distribution tends to be affected as well. This rearrangement of pore geometry by roots is likely to lead to significant changes in processes such as water and nutrient uptake (Daly et al. 2015).

There is a tendency for soil porosity to increase as it approaches the root surface, as shown in Figure 7 (Kuzyakov and Razavi 2019; Burr-Hersey et al. 2020). Helliwell et al. (2017) observed that above 1 mm from the root surface, the porosity decreases to the porosity value of the non-rhizospheric soil

(Figure 7). The presence of larger pores in the rhizosphere soil is reported due to root growth, but there is also an increase in micropores around roots associated with microorganisms (Whalley et al. 2005; Feeney et al. 2006). It is also known that sandy soils tend to show a greater increase in rhizosphere soil porosity than clayey soils, and that changes in rhizosphere soil porosity occur over time, as shown in Figure 7 (Helliwell et al. 2017).

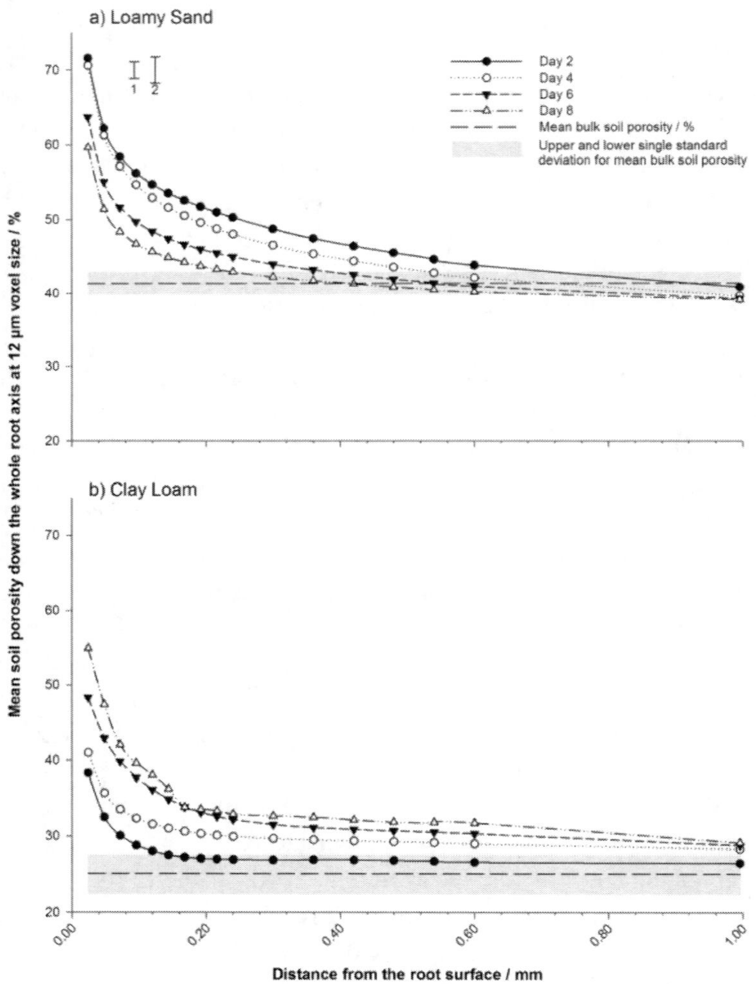

Source: Adapted from Helliwell et al. (2017).

Figure 7. Mean soil porosity for (a) loamy sand, and (b) clay loam soils from the root surface over time. Standard errors of the difference are shown for (1) sampling day and (2) soil texture.

Mucilage is associated with changes in the soil pore network of the rhizosphere. It partially adheres to the soil matrix, forming a polymeric network, and is hydrophobic when dry (Kroener et al. 2014; 2018). This suggests that the mucilage could create pores in the soil of the rhizosphere when dry, which would not be present when wetted. In addition, mucilage expands during wetting, increasing its volume (Czarnes et al. 2000). This expansion may be related to the increase in porosity of the rhizosphere soil, as this expansion should displace soil aggregates away, as occurs during root growth.

The rhizosphere soil also changes its hydraulic conductivity relative to the non-rhizosphere soil. Initially, it was thought that the increase in porosity could lead to a greater hydraulic conductivity of the rhizosphere soil. However, the effects of mucilage overlap. Thus, the hydraulic conductivity in the presence of mucilage is no longer dependent on the soil particle size because the mucilage is partially bound to the soil matrix, creating a polymeric network that increases the resistance to water flow in the pores (Kroener et al. 2018).

Therefore, the amount of mucilage exuded by the root is directly related to hydraulic conductivity. An increase in this amount leads to an increase in the viscosity of the liquid phase and a decrease in the wettability of water molecules, thus reducing the hydraulic conductivity (Kroener et al. 2014; 2016). To get an idea, the viscosity of mucilage at a concentration of 0.5% (g dry mucilage g^{-1} water) is three times higher than that of pure water (Ahmed et al. 2017). Thus, after exudation, the mucilage may be able to reduce the hydraulic conductivity of the rhizosphere soil, thereby facilitating root water uptake for at least one day (Ahmed et al. 2014). Therefore, when the rhizosphere soil and the soil outside the rhizosphere are maintained at the same water content, a reduction in hydraulic conductivity in the rhizosphere soil is observed due to the presence of mucilage (Kroener et al. 2014).

Some changes in soil properties in the rhizosphere occur as a plant response to the reduction in soil hydraulic conductivity in this environment, among other factors mentioned above (Benard et al. 2019). Kroener et al. (2018) developed a conceptual model of the interactions between soil particles, mucilage, and water and how they affect hydraulic conductivity in the saturated state (Figure 8). In addition, mucilage increases the viscosity of the soil solution and, when present in large quantities, can exert a strong resistance to the water flow in soil pores (Ahmed et al. 2017; Kroener et al. 2018). This mucilage affects the surface tension of the liquid and its contact angle with the solid, altering the meniscus and consequently the flow rate inside the pore containing the mucilage, as shown in Figure 9 (Whalley et al. 2005; Kroener et al. 2018).

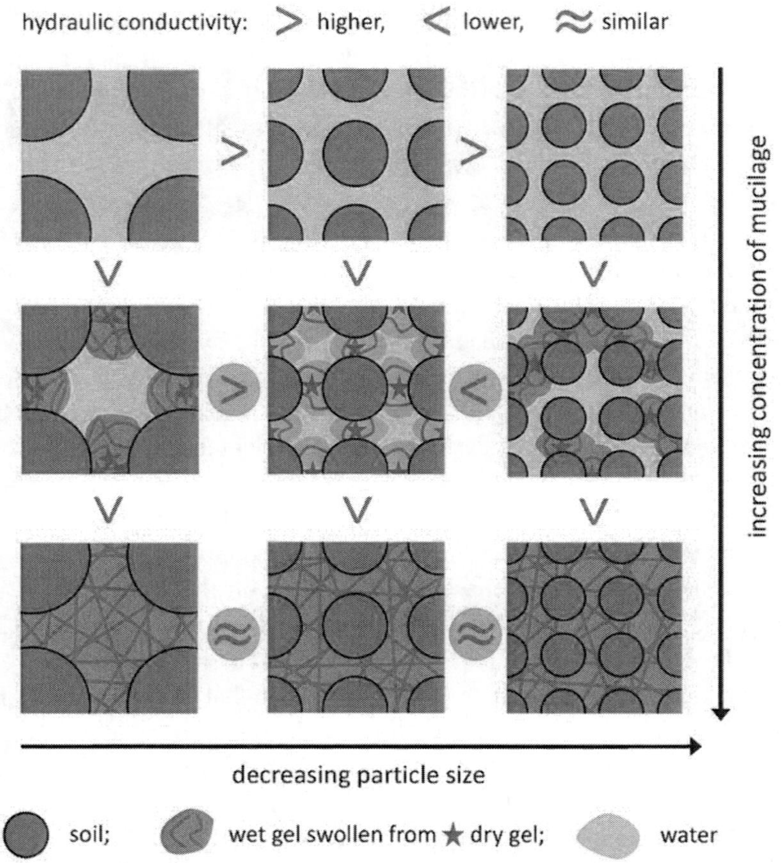

Source: Adapted from Kroener et al. (2018).

Figure 8. Conceptual model of the saturated hydraulic conductivity as a function of particle size and mucilage concentration. In each row, the concentration of mucilage per unit of soil volume is constant for all soils. The positions of the stars indicate the place where mucilage is deposited on the particle surface as the sample dry.

Differences in soil water retention are also observed in the rhizosphere of different plants (Whalley et al. 2005). According to these authors, these differences are due to the different composition of the mucilage exuded by different plants. On the one hand, some authors report a higher content of retained water in non-rhizosphere soil than in rhizosphere soil (Whalley et al. 2005; Daly et al. 2015). On the other hand, mucilage is thought to be able to absorb large amounts of water at any water content, thereby increasing the water holding capacity of the soil (Kroener et al. 2014; Ghezzehei and Albalasmeh 2015). Thus, it is expected that rhizosphere soils are often associated with higher water contents than non-rhizosphere soils. However, there is no single relationship between water content and mucilage potential of rhizosphere soil, and this relationship varies temporally with soil wetting and drying cycles (Kroener et al. 2014). According to these authors, mucilage increases the water content of the rhizosphere soil when it is in equilibrium with the matric potential, but if not rehydrated, it can decrease the water content of the rhizosphere.

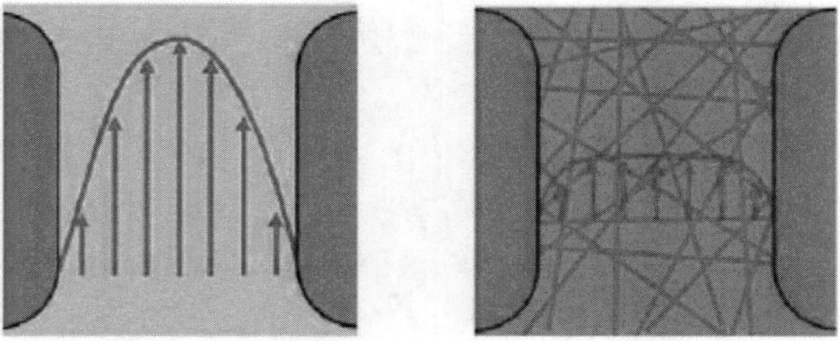

Source: Adapted from Kroener et al. (2018).

Figure 9. Conceptual model of the flow velocity in a pore without mucilage (parabolic) and in a pore filled with mucilage (nearly uniform).

Neutron radiography has been used to visualize water movement in the rhizosphere. Carminati et al. (2010) first used this technique to monitor water movement in the rhizosphere soil. Later, it was adapted to quantify water fluxes and measure which region of the roots absorbs water from the soil using deuterated water (Zarebanadkouki et al. 2012). This methodology was validated by further studies, which showed that the water content of the rhizosphere soil is significantly different from the bulk soil (Ahmed et al. 2014; 2016; Zarebanadkouki et al. 2014; 2018). These studies revealed that during soil drying, the mucilage allows the rhizosphere to remain moist for a longer period, and when the soil is wetted, the mucilage rehydrates at a lower rate than the soil, temporarily reducing soil rehydration from the rhizosphere.

However, several factors need to be considered. For example, this rate of rehydration or rewetting is related to the age of the rhizosphere. Young roots are covered by fresh and hydrated mucilage, which helps to capture water and other resources, while the old roots are also covered by older mucilage, which becomes water repellent as it is rehydrates more slowly (Carminati and Vetterlein 2013). This relationship is illustrated in Figure 10. Furthermore, it should be noted that it is not possible to correctly describe the dynamics of water in the rhizosphere soil with the classical water retention curve, in which the relationship between water content and matric potential is unique (Kroener et al. 2014).

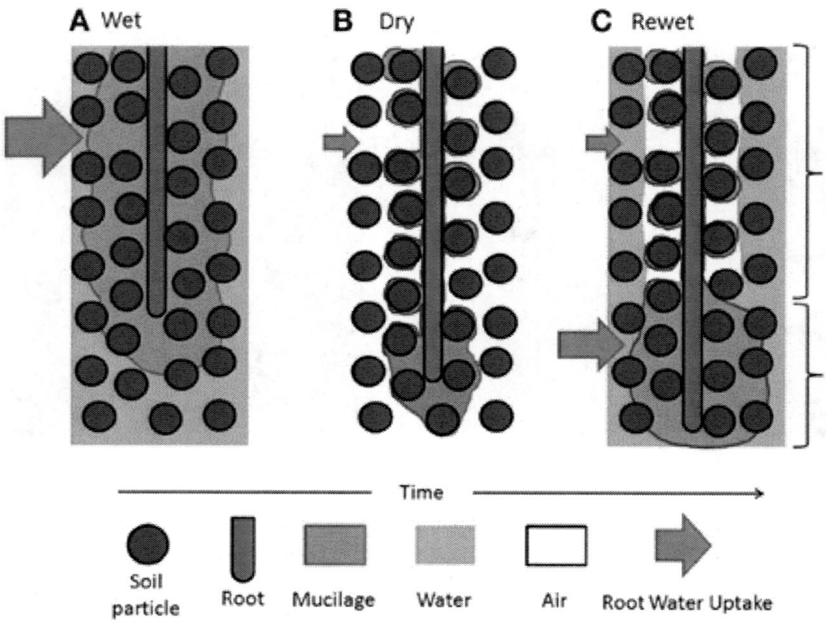

Source: Adapted from Carminati (2013).

Figure 10. Conceptual model of water and mucilage distribution in the rhizosphere during (a) wetting, (b) drying and (c) rewetting cycles.

Probably the most important factor reflecting changes in the physical and hydrological properties of the rhizosphere soil is mucilage (Carminati et al. 2010, 2016; Zarebanadkouki et al. 2016). Thus, mucilage extracted from chia seeds (*Salvia hispanica*), which has a similar composition to root exudates, especially from plants such as lupine and maize, has been used as a good surrogate in studies to simulate the effects of root mucilage in soil (Carminati and Vetterlein 2013; Naveed et al. 2017; van Veelen et al. 2018). Therefore, studies on soil stability and dynamics of water and herbicides in soil have been developed using this mucilage extracted from chia seeds (Kroener et al. 2014;

Benard et al. 2016; van Veelen et al. 2018; Zhong et al. 2021). Exudates can also be extracted directly from plant roots (Ahmed et al. 2016; Giles et al. 2017; Naveed et al. 2017). Advances such as these have been very important in expanding our knowledge of the influence of roots and their rhizosphere on the soil. This is because it has become possible to isolate a factor responsible for changes in the rhizosphere and to study it controlling its proportion and the other factors involved.

Managing the Rhizosphere

A better understanding and control of rhizosphere processes is one of the most important scientific tasks of our time, given the global challenges of climate change and population growth (McNear 2013). Many of the rhizosphere processes discussed above are related to crop production and environmental conservation.

Studies of changes in the rhizosphere indicate the importance of knowing the response of different plant species in different soils, since organic matter release occurs in response to the environment, and the choice of crops to be grown is extremely important in soil conservation planning. The knowledge of the influence of the roots on the soil can help in the decision of the crop rotation system to be adopted by the producer (Batista et al. 2020).

Different plant roots can result in different "rhizosphere effects" leading to changes in microbial populations and electrochemical components (Li et al. 2021; Batista et al. 2022; Wang et al. 2022). These changes can lead to alterations in important biogeochemical cycles. Changes in soil aggregation are also mediated by the rhizosphere and depend on the environment in which the plant is inserted, e.g., the same plant in soils with different properties and/or characteristics or in the same soil under different management presents different exudation, modulating its rhizosphere according to the environment, in this case the soil. Therefore, further studies are needed to understand the complexity of the universe around the roots, the soil rhizosphere, and to use this knowledge to help define and implement conservation agriculture practices, especially regarding the choice of cultivars that promote greater benefits to the soil due to the substances released by the roots.

Once the physical and hydraulic properties of the soil in the rhizosphere are altered, knowledge of these changes can lead to improvements in water retention in the rhizosphere soil, resulting in better water uptake by plants. This can lead to higher productivity, for example, if additional water uptake during the grain filling period increases grain yield, which is of economic interest (Ahmed et al. 2018).

Advanced technologies to determine rhizosphere soil porosity, allowing 3D visual assessments, have provided more robust information about the pore network of this soil. However, as with other physical and water properties, changes in rhizosphere soil porosity depend on the crop being grown, with a strong influence of root architecture, which is influenced by soil texture (Burr-Hersey et al. 2020; Helliwell et al. 2017). These authors also suggest that the alteration of soil pore space in the rhizosphere may act as a self-remediation of plants growing in compacted soils through structural reorganization.

Understanding the water dynamics in the rhizosphere soil can help guide management practices aimed at improving water and nutrient uptake by plants. For example, Zarebanadkouki et al. (2018) were able to reduce water repellency in the rhizosphere soil of white lupine (*Lupinus albus* L.) after wetting with a commercial surfactant, thereby increasing the rehydration rate.

The response of root exudates as a function of drought-induced changes was investigated by Vries et al. (2019) in two common temperate species: *Holcus lanatus* and *Rumex acetosa*. The authors observed that C cycling was affected by changes in respiration, which was influenced by the production of root exudates. Another relevant aspect was that the root exudates of these plants showed an ability to cycle the C of the ecosystem, due to the C decomposition from the soil. These aspects showed a certain evolutionary strategy of the studied plants to overcome adverse environmental conditions, by stimulating microbial activity and mineralization of nutrients. Baumert et al. (2018) investigated the importance of natural root exudation from a European beech (*Fagus sylvatica* L.) forest on C allocation and aggregation in the topsoil and subsoil. A model root exudate was experimentally evaluated at two different concentrations using artificial roots. The addition of exudate favored fungi in the microbial community composition, which consequently increased the formation of aggregates > 250 μm due to the greater root exudation.

Soil pathogens pose a significant challenge to agricultural production due to the limitations of control by conventional agronomic practices (Wille et al. 2019; Akamatsu et al. 2020). It is estimated that 20-30% of agricultural productivity is lost due to soil-borne pathogens worldwide (Savary et al. 2019; Vannier et al. 2019). The Food and Agriculture Organization of the United Nations (FAO) estimates that plant diseases cost the global economy around US$220 billion annually (FAO 2021). Most agricultural soils are infested with some level of plant pathogens; however, their presence does not necessarily result in plant infection (Gu et al. 2022). Therefore, understanding the role of the rhizosphere microbial community in controlling plant root diseases is considered one of the most important research fronts to increase productivity and confer sustainability to agricultural practices.

The advent of advanced molecular techniques has contributed to a better understanding and prediction of the multitrophic ecological interactions between the components of the classical pathosystem triangle, composed of hosts, pathogens, and the environment (Francl 2001; Frantzeskakis et al. 2020). Thus, from a holistic perspective, it is possible to add the dimension of the rhizosphere microbiome to the pathosystem triangle, where the environment (soil) is the main source of microorganisms. In this system, the host plant recruits soil microorganisms to the rhizosphere through the exudation of signaling compounds. The pathogen competes with the complex microbial community of the rhizosphere, aiming to reach the plant root system and subsequently infect and cause plant disease (Mendes et al. 2013; Kwak et al. 2018; Carrion et al. 2019).

A better understanding of these interactions was the basis for the description of important definitions. Plant disease suppressive soils refer to the condition in which soil biological suppressiveness can contribute to the reduction or inhibition of the root pathogen establishment and/or persistence by the presence of organisms in the rhizosphere that are antagonistic to the disease (Baker and Cook 1974; Mendes et al. 2011; Raaijmakers and Mazzola 2016). Soil biological suppressiveness is attributed to the activity of the microbial community and can be divided into two types: general suppressiveness, which is related to the general mechanisms of competition, parasitism and antibiosis between microorganisms and the pathogen and specific suppressiveness, which is related to the activity of specific groups of microorganisms that interfere at some stage of the pathogen life cycle (Expósito et al. 2017).

The plant rhizosphere is formed by the selection of microorganisms from the soil through the exudation of biochemical compounds (Huang et al. 2014; Venturi and Keel 2016; Robertson-Albertyn et al. 2017). However, the composition of the rhizosphere microbiome can be influenced by soil pathogens attacks through the activation of plant defense mechanisms. These mechanisms modulate the assembly of the rhizosphere microbial community through the exudation of specific compounds, aiming to select soil microorganisms for protection against the invader (Mendes et al. 2011; Afridi et al. 2022). For example, Lebeis et al. (2015) found that the phytohormone salicylic acid modulates the colonization of Arabidopsis roots by specific bacterial taxa.

Thus, the pathogen success is directly linked to its interaction with the rhizosphere microbiome (Wei et al. 2015). Several studies have considered that the rhizosphere microbiome is the first line of defense of the plant against attack by soil pathogens (Trivedi et al. 2017; Carrion et al. 2018; Durán et al. 2018). Some studies correlate greater efficiency of disease suppression with the increase of soil biodiversity (Bender et al. 2016; Saleem et al. 2019), where greater interaction between the community favors a high competition for niches, resulting in greater resistance to pathogen invasion (Wei et al. 2015; Mendes et al. 2018). However, pathogen infection can also result in an increase or decrease in the diversity of the rhizosphere microbiome (Liu et al. 2016; Wei et al. 2018; Lee et al. 2021).

Several studies have reported that root exudates released under successive monocultures increase pathogen accumulation (Wu et al. 2019; Li et al. 2020). Reduced biodiversity in the system may reduce competition for resources between the pathogens and the microbial community, favoring the development of pathogens that must invade the root microbiome and multiply as saprophytes until they reach a population density to infect and cause disease (Wei et al. 2019). However, one study has demonstrated that changes in the composition of the rhizosphere can explain future plant health. Yian et al. (2022) assessed the relationship between the composition of the tomato rhizosphere microbiome and disease caused by the bacterium *Ralstonia solanacearum* and verified that changes in the composition of the rhizosphere microbiome could predict disease onset up to two weeks before detectable changes in pathogen abundance.

Understanding how these complex multitrophic ecological interactions occur in the rhizosphere and influence their functions, resulting in the decrease or inhibition of plant disease-causing pathogens is a major challenge to optimize and improve the manipulation of the rhizosphere microbiome (Afridi et al. 2022). Rhizosphere microbiome engineering is an emerging tool that aims to manipulate the host or its associated microbiome to optimize functions of interest. Therefore, this resource can be explored to improve plant growth and assist in the control of soil-transmitted diseases (Mukhtar et al. 2018; Pérez-Jaramillo et al. 2018).

Disease-suppressing microbiomes may be related to the enrichment of specific microbial taxa, such as Pseudomonadota, a phylum associated with the health of the rhizosphere (Mendes and Raaijmakers 2015; Dhaoaudi et al. 2018; Gu et al. 2022). Thus, the identification and isolation of these taxa is an interesting strategy for disease suppression; however, most of these taxa cannot be recovered by standard isolation procedures. The rhizosphere microbiome can contain 10 to 100 times more microorganisms than the soil (Drogue et al. 2013; Kumar et al. 2019), which has an average abundance of live cells between 10^7 and 10^9 per gram of soil (Alexander 1977). Therefore, the rhizosphere microbiome is more efficient at inhibiting pathogens than a single strain (Kumar and Verma 2019).

The inoculation of microorganisms as biological control agents (BCAs) in the soil, seeds, and planting materials is a consolidated strategy and has recently received greater scientific and commercial attention as a promising alternative or complementary strategy for the management of soil pathogens (Niu et al. 2020; Dimkic et al. 2022). The increased use of BCAs is associated with the stability, economic benefits, long-term effects, security, and alignment with the principles of sustainable agriculture that this strategy offers (Vazquez et al. 2000; Filo et al. 2013).

The prospection of rhizosphere microorganisms has been widely explored to obtain BCAs with potential for the biocontrol of plant root pathogens (Suprapta 2012). A promising strategy has been the prospection of microorganisms from the plant rhizosphere in disease-suppressing soils (Mendes et al. 2011). Several works show the efficacy of these microorganisms in biocontrol against pathogen invasion through the production of antibiotics and production of hydrolytic enzymes, volatile organic compounds inhibitors of pathogens, and competition in the rhizosphere and siderophores (Hopkins et al. 2017; Berg et al. 2018; Prajapati et al. 2020). A wide range of species and genera of microorganisms can be cited *Agrobacterium*, *Arthrobacter*, *Bacillus*, *Burkholderia*, *Collimonas*, *Enterobacter*, *Paenibacillus*, *Pedobacter*, *Pseudomonas*, *Rahnella*, *Serratia*, *Lysobactyer*, and *Trichoderma* (Schlatter et al. 2017; Ali et al. 2022b).

However, the number and type of strains that should be used to achieve greater efficiency is not well defined. It is known that the more diverse the consortium, the greater the chance of establishment and survival in the rhizosphere microbiome, increasing resistance to pathogen invasion and plant infestation (Bell et al. 2005; Hu et al. 2016; Kristensen et al. 2016). Inoculum diversity can boost the functionality of bacterial communities in the rhizosphere through intensified competition for resources between the introduced bacterial community and a potential pathogen, competition for interference with other microorganisms, including resident microbiota and pathogens (Van Elsas et al. 2012; Wei et al. 2015). For example, a combination of different bacterial secondary metabolites produced by consortium diversity may result in stronger antagonism against the pathogen (Loper et al. 2012).

Finally, we can consider that the reduction or inhibition of plant diseases caused by root pathogens, resulting from the advances in knowledge between the interactions of the rhizospheric microbiome, host plant, pathogen and the environment, has effectively contributed to the definition of more sustainable management strategies. However, it is important to highlight the magnitude of this challenge, which is to understand the complex and sensitive ecological interactions between the components of the pathosystem triangle with the soil microbiome and/or rhizosphere through the range of factors that can influence the pathogen infection process in the host plant root and disease development, such as plant genetics, nature/physiology/ecology of the pathogen, differences in soil management, environmental conditions, and interactions between the exogenous BCAs inoculated with the existing microbiota and the host plant.

Root development, rhizosphere-mediated processes, and nutrient cycling can work synergistically to facilitate processes that enhance plant growth in multicropping systems and help suppress soil-borne diseases, pests, and weeds (Ehrmann and Ritz 2014). As shown by these authors, ecological, biochemical, and physical mechanisms are involved in the natural defense mechanisms of plants (Figure 11).

Agroforestry systems (AFS) favor the maintenance of important ecosystem services when compared to monoculture systems such as shown in Figure 12, higher soil organic carbon (SOC) stocks, pest control, water quality, water cycling, nutrient cycling, heavy metal removal, system biodiversity, and higher yields (De Stefano and Jacobson 2018; Lovell et al. 2018; Schmidt et al. 2022; Bian et al. 2022). The greater potential to access and absorb nutrients from tree species in deep layers favors a greater SOC stocks, greater porosity, and nutrient cycling through litter deposition (Cong et al. 2015; Isaac and Borden 2019). Biodiversity is increased in the AFS, as a result of the greater diversity of litter deposited and exuded in the rhizosphere, which contributes to aboveground and belowground diversity (Eisenhauer et al. 2017; Chen et al. 2019a).

AFSs can affect the diversity and structure of microbial communities in the rhizosphere, allowing for new beneficial interactions (Araujo et al. 2012; Liang et al. 2020). Higher root biomass of perennial vegetation promotes greater association of microorganisms with roots, for example, mycorrhizal fungi, symbiotic bacteria, and associative bacteria, through increased rhizodeposition and easier access to newly assimilated carbon (Pudjiasmanto et al. 2018; Chaiya et al. 2021). This effect can also be attributed to the greater diversity of exudates in the rhizosphere (Prescott and Grayston 2013; Beule et al. 2020), which can change both soil pH and the levels of nutrients available in the rhizosphere (Duchene et al. 2017; Liu et al. 2019), such as available nitrogen (Deng et al. 2022), phosphorus and available potassium (Zhang et al. 2018).

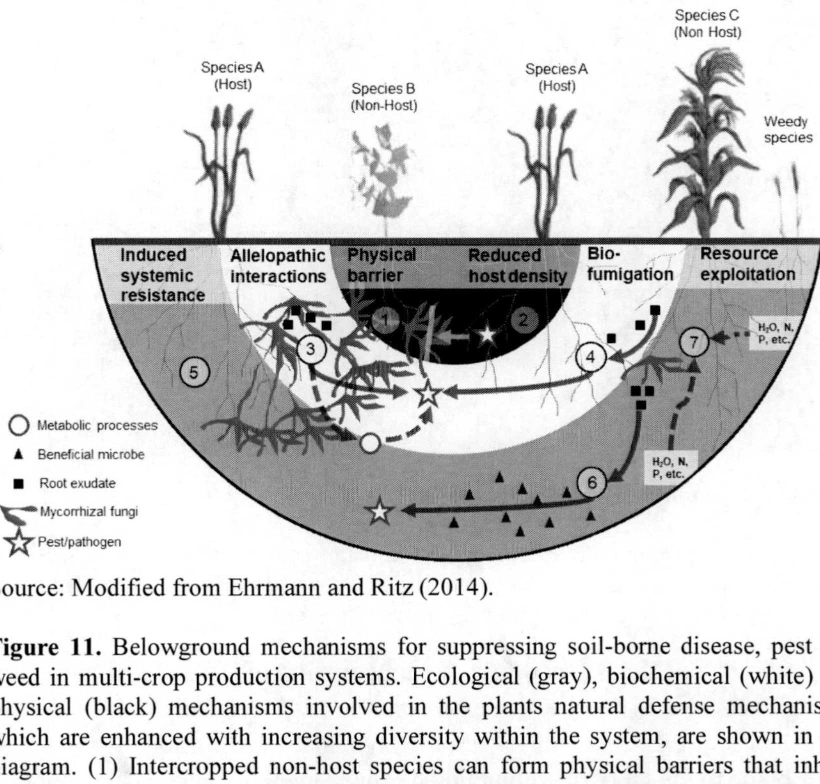

Source: Modified from Ehrmann and Ritz (2014).

Figure 11. Belowground mechanisms for suppressing soil-borne disease, pest and weed in multi-crop production systems. Ecological (gray), biochemical (white) and physical (black) mechanisms involved in the plants natural defense mechanisms, which are enhanced with increasing diversity within the system, are shown in this diagram. (1) Intercropped non-host species can form physical barriers that inhibit pathogens and pest from infesting neighboring host plants. (2) Intercropping reduces the host plant density, which significantly reduces the severity of disease/pest infestations; (3) Roots exude allelochemicals that are converted directly or through microbial metabolic processes to inhibit the growth of selective other plant species, and pathogenic microbes. (4) Biofumigant compounds, e.g., from brassicas such as glucosilonates are transformed to isothiocyanides and nitriles. (5) Pathogen attack and mycorrhizal associations can induce systemic resistance, and enhance the plants' own natural defense mechanisms. (6) Increased amounts and types of root exudates attract beneficial microbes, including natural enemies, parasites, and pathogens. (7) Resource complementarity and more efficient use of resource in intercrops can suppress weed infestations.

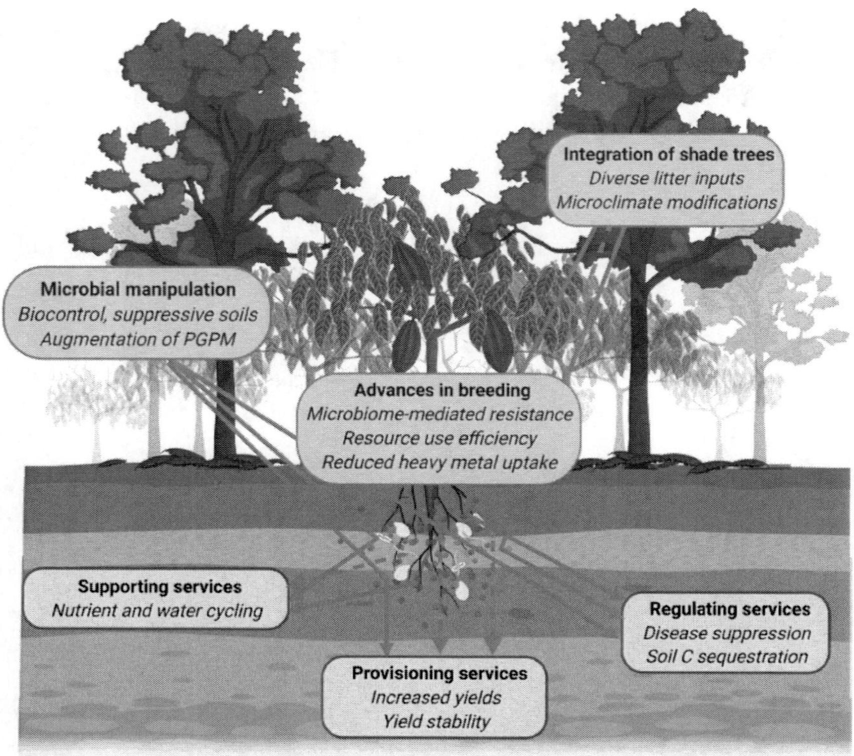

Source: Adapted from Schmidt et al. 2022.

Figure 12. Rhizobium-targeted strategies to enhance ecosystem services in AFS. Solid arrows represent direct effects and dashed arrows represent indirect effects. C, carbon; PGPM, plant growth-promoting microorganisms.

Deng et al. (2022) found that the implantation of AFS in a monoculture of T. grandis altered the chemical properties of the rhizosphere soil, in addition to promoting changes in bacterial functional structuring, where AFS contributed to the increase of functional groups involved in soil C metabolism and decreased the functional groups involved in soil N metabolism.

Leguminous are highly valued for the ecosystem services they provide, being an excellent alternative for diversification in the AFS field (Apolinário et al. 2015; Isaac and Borden 2019). Among the main benefits are the increase in nitrogen levels in the soil through direct pathways, root exudates rich in N, mycorrhizal networks, or indirect pathways such as higher quality litter decomposition, mining processes, and biological nitrogen fixation by the symbiotic association between diazotrophs and leguminous (Isaac and Borden

2019; Lira Junior et al. 2020; Wang et al. 2021). Previous studies have shown that leguminous when in intercropping systems can be forced to fix more atmospheric N_2, due to competition for inorganic nitrogen from the soil (Hauggaard-Nielsen et al. 2003; Neumann et al. 2007).

Barros et al. (2018; 2021) found that the introduction of legumes in a silvopastoral system increased the abundance of total bacteria, ammonium oxidizing bacteria, and diazotrophic bacteria. According to the authors, these results can be attributed to the deposition of organic compounds via exudate, improving soil physicochemical properties (Moura et al. 2015) and resulting in favorable microhabitats, in addition to the exudation of chemical signaling compounds in the rhizosphere by the legumes, stimulating N-fixing bacteria (Janczarek et al. 2015).

The introduction of leguminous in AFS also contributes to the P cycle (Rodrigues et al. 2021). Plants and microorganisms, alone or in the consortium, can exude low molecular weight organic acids, phytases, and phosphatases in the rhizosphere that acidify the pH and contribute to the solubilization of calcium, iron, and aluminum phosphate, in addition to organic P mineralization (Erel et al. 2017; Sun et al. 2020). George et al. (2002) found that phosphatase activity increased significantly with the proximity of the rhizosphere and with the growth of agroforestry species. Liu et al. (2021) evaluated the short-term impact of the introduction of a perennial legume (*Flemingia macrophylla*) in two AFS (young and mature), where it was verified that the introduction of the leguminous as a cover crop in AFS contributed to the cycling and availability of P in the rhizosphere, mainly in young AFS due to abundant return of leguminous residues.

Implementation of AFS may be a promising management strategy for the phytoremediation of heavy metals. Bian et al. (2022) evaluated the remediation potential of Lei bamboo (*Phyllostachys praecox*) in pure plantation and agroforestry systems in soil contaminated with heavy metals (Cd/Zn/Cu). They found that bamboo roots removed greater amounts of heavy metals from the rhizosphere soil, highlighting the greater removal of Cu and Cd in AFS. According to the authors, rhizosphere SOM is a crucial factor affecting the immobilization of heavy metals by adsorption, complexation, or chelation (Wen et al. 2018) and the absorption of heavy metals adsorbed in rhizosphere SOM by bamboo is the main remediation pathway of heavy metals (Nath et al. 2015).

It is important to highlight that the role of the rhizosphere microbiome in AFS deserves further study. Especially considering the longer lifespan of perennials and the consistent presence of roots throughout the year making rhizosphere interactions more durable in tree crops than in annual crops (Mercado-Blanco et al. 2018). Thus, for a better understanding of the ecological interactions between AFS and the rhizosphere microbiome some key questions are: "How does the greater diversification induced by the implantation of AFS relate to the functions performed by the rhizosphere microbiome?"; "How can AFS diversification promote the development of disease-suppressing soils?"; "Which microbial taxa make up the core microbiome in AFS?"; and "What are the relevant ecological roles and functions for plants of central and variable taxa?".

Finally, it is important to highlight that the benefits of AFS implantation concerning monocultures are well-reported in the literature for several crops (Munroe et al. 2015; Zhang et al. 2018; Liu et al. 2021; Bian et al. 2022; Deng et al. 2022; Schmidt et al. 2022; Zhou et al. 2022). Furthermore, the implementation of AFS is an underexplored strategy given its potential to address global environmental challenges, for example, mitigation and adaptation to climate change (Hernández-Morcillo et al. 2018), preservation and promotion of biodiversity (Gama-Rodrigues et al. 2021), improve soil health (Dollinger and Jose 2018), increase farm income (Staton et al. 2022), and thereby strengthen environmental, economic, and social benefits (Elevitch et al. 2018).

Modern agriculture seeks strategies to increase crop yields due to the production stagnation that has occurred in recent years. Although efforts have been made to improve the chemical properties of soil, greater attention has been focused on the biological and physical sectors in order to strengthen the eco-friendly strategies. The major bottleneck is to understand how plants and the autochthonous microbial community would respond to the use of these biostrategies. In addition, it is crucial to understand how application exogenous microorganisms in the soil, as well as the application of organic compound would affect the natural soil microbial community and plant response (Mawarda et al. 2020). Therefore, the success of biological management depends on the understanding of the plethora of microbes that inhabit the rhizosphere alongside the other physical and chemical parameters.

It is impossible to explain soil microbes without mentioning their importance in maintaining soil health. Healthy soil can perform its chemical, physical and biological functions working together, providing food production for the maintenance of human existence. A better understanding of soil health comes from its analogy to the human body, that is, we define a healthy person by their ability to meet their daily needs. Likewise, soil has many functions that go beyond supporting food production, including providing water and air, physical support, the source of minerals used in technology, the main source for mitigating climate change, and so on. In other words, if we had to pay for these ecosystem services mentioned earlier, our monthly bill would be much more expensive.

One of the main ecosystem services provided by rhizosphere microbes that stood out among others is undoubtedly the biological nitrogen fixation. In the Brazilian context, it represents an economic saving of more than US$ 13 billion, with 78% of the soybean cropping area (approximately 37 million hectares) being inoculated annually with bacteria of the genus *Bradyrhizobium* (Santos, Nogueira, and Hungria 2019). However, recent research has shown that the use of mixed inoculants is promising, mainly using bacteria of the genus *Bradyrhizobium,* combining with *Azospirillum,* and *Trichoderma*, resulting in grain yield of over 14% (Barbosa et al. 2021; Barbosa et al. 2022).

In addition, rhizobacteria inoculation in sugarcane combined with the organic compost application has been shown to be a promising eco-friendly strategy to increase sugarcane yield up to 6% (Lopes et al. 2021). This is due to subtle changes in the composition and function of the microbial community (Silva et al., 2022). Furthermore, according to Silva et al. (2023), soil enzyme activities related to phosphorus and carbon cycling can be used as predictors of sugarcane yield when rhizobacteria was inoculated and, therefore, generating feasible metrics that can be used by farmers to predict the productivity and equivalent economic gains. In addition, management of rhizosphere based on inoculation of phosphate-solubilizing bacteria is being well highlighted in different maize genotypes, including the increase in yield and quality parameters (Pardo-Díaz et al. 2021; Beltran-Medina et al. 2022).

As we mentioned at the beginning of this chapter, the microbial community of the rhizosphere is inhabited by a myriad of microbes, and therefore, their inoculation-based management is not restricted to bacteria. Inoculation with arbuscular mycorrhizal fungi (AMF) is another sustainable strategy to increase plant biomass as well as alleviate the stress caused by abiotic and biotic conditions (Jongen et al. 2022; Yin et al. 2022; Zeng et al. 2022). Several studies have shown that AMF combined with rhizobacteria have their results potentiated due to the synergism between them (Jiang et al. 2021; An et al. 2022). However, it is crucial to understand how the resident microbial community would respond to the introduction of these new microbes. In this regard, Mawarda et al. (2020) published a detailed review, showing the possible ways that the autochthonous microbial community would respond to the inoculated microbes. It is well known that the time of persistence of some inoculants in the soil varies significantly, but overall they did not persist for long in the soil, however they can cause changes in the resident microbial community.

According to Mawarda et al. (2020), when a microbial inoculant is introduced into a soil with a resident microbial community, first, they interact with each other, but the resident community may retain function and structure regardless of the amount of inoculum added, and therefore, exhibit a trait of resistance (resistant) (Figure 13a). Second, the resident microbial community may show the ability to self-organize after disturbance, returning to its original state, thus showing a trace of resilience (transient impact) (Figure 13b). Third, after the interaction of the resident community and the inoculant, an alternative stable state may be achieved, and therefore, the shift of the resident community remains (non-transient impact) (Figure 13c). Therefore, the success of the rhizosphere management based on microbial inoculants must consider the genotypes of plant species, the edaphoclimatic conditions, the inoculant traits, and the composition of the resident microbial community.

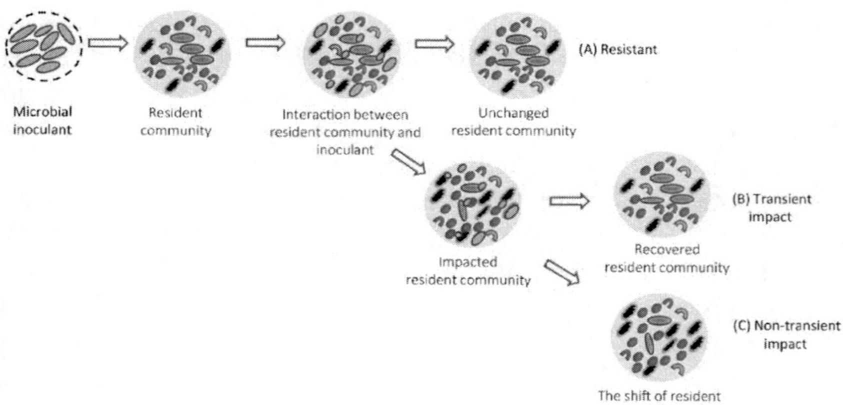

Source: Adapted from Mawarda et al. (2020).

Figure 13. Possible responses of the resident microbial community to inoculant input into the soil. (a) resistance traits, (b) transient impact, and (c) non-transient impact.

In addition, inoculant investigations should consider a wide range of host plants, as some plants would respond differently to microbes. The most proeminent example is the use of *Arabidopsis* plants to reveal important molecular and biochemical responses. However, *Arabidopsis* will not be helpful in understanding mycorrhizal symbiosis because it is a genus of the family *Brassicaceae*, known as a non-mycorrhizal family.

Furthermore, it is important to consider the impact of the COVID-19 pandemic on the microbial dynamics of the rhizosphere and the enormous need to fortify our crops (biofortification process), since human immunity depends on our nutritional status (El-Ramady et al. 2021). In addition, it is beneficial to maintain soil-plant-human health as an integrated system, reinforcing the design of cutting-edge projects to fill this gap, where medical doctors will work together with farmers, agronomists, biologists and scientists. Therefore, the main challenge is to translate scientific findings into practical real-world applications that start in the field and travel to the human body (Jones and Hinsinger 2008; Philippot et al. 2013).

Going further, as we did for the soil chemical recommendation, we need to gather knowledge regarding biological and physical fertility of different soil types, locations, and types of crops/plant species, among others. This is crucial to building soil health. In addition, one of the great challenges for new management strategies is to consider the current trilemma, that is, climate change, food security and environmental degradation (Cerri et al. 2018). As scientists working for society, we need to focus our efforts on disseminating all the scientific information that we already have, but this also requires a political effort on our favor, ensuring that the knowledge generated reaches farmers. We need to encourage them to consider soil as a living component of the system, and therefore encourage them to follow sustainable management practices.

Conclusion

Rhizosphere is the most active part of the soil where soil properties change in space and time. The rhizodeposition process is responsible for the selection of the microbial population with higher activity in this region. This process depends on several factors (i.e., soil biota, physiological stage of the plant, environmental conditions and substrate properties). The substances released by the plant roots act as the main driving force in the rhizosphere, leading to the release of elements in this soil region and altering chemical parameters. Specific biogeochemical processes take place in this region of the soil due to the plant-soil-biota relationship, enhancing processes such as weathering and soil formation. The rhizosphere also plays an important role in the exchange of elements and energy flow. Enzymes from plants or microorganisms can also release nutrients or make them available to plants. They indicate microbial activity and are sensitive indicators of changes in soil ecology.

Beneficial microorganisms in the rhizosphere can act as biocontrols, promote plant growth or be useful as biofertilizers, once microorganisms participate in serious biogeochemical cycles of elements. However, physical parameters are less dynamic than biological or chemical parameters, they also change in the rhizosphere, mainly due to root exudates and mucilage, both released by the roots.

Thus, rhizosphere management can help improve crop production while maintaining soil sustainability once nutrients and water absorbed by plants have passed through it. Rhizosphere management is useful to control pathogens and improve plant development. Understanding the process and factors that act in the rhizosphere for different types of plants and soils is useful in selecting specific crops or intercrops or adopting some management practices to increase soil health and crop production.

In theory, the rhizosphere it is easy to define, but the relationship between plant soil and biota is one of the most complex experienced by terrestrial plants, and it can change in space and time with numerous factors. Thus, the assessment of the rhizosphere is very difficult due to methodological limitations, especially for in situ measurements.

References

Abbott, A. 2016. Scientists bust myth that our bodies have more bacteria than human cells. *Nature* 10:1-2.

Afridi, M. S., Fakhar, A., Kumar, A., Ali, S., Medeiros, F. H., Muneer, M. A., Ali, H., Saleem, M. 2022. Harnessing microbial multitrophic interactions for rhizosphere microbiome engineering. *Microbiological Research* 265: 127199.

Ahmed, M. A., Kroener, E., Benard, P., Zarebanadkouki, M., Kaestner, A., Carminati, A. 2016. Drying of mucilage causes water repellency in the rhizosphere of maize: Measurements and modeling. *Plant and Soil* 40:161-171.

Ahmed, M. A., Passioura, J., Carminat, A. 2018. Hydraulic processes in roots and the rhizosphere pertinent to increasing yield of water-limited grain crops: A critical review. *Plant and Soil*, 69:3255-3265.

Ahmed, M. A., Kroener, E., Holz, M., Zarebanadkouki, M., Carminati, A. 2014. Mucilage exudation facilitates root water uptake in dry soils. *Geophysical Research Abstracts* 16: EGU2014-2978.

Ahmed, M. A., Zarebanadkouki, M., Ahmadi, K., Kroener, E., Kostka, S., Kaestner, A., Carminati, A. 2017. Engineering rhizosphere hydraulics: Pathways to improve plant adaptation to drought. *Vadose Zone Journal* 17:160090.

Akamatsu, H., Fujii, N., Saito, T., Sayama, A., Matsuda, H., Kato, M., Kowada, R., Yasuta, Y., Igarashi, Y., Komori, H., Tanji, K., Kuroda, T., Fujita, Y., Hattori, M., Kawakami,

O., Hori, T., Mimuro, G., Morikawa, T., Murasaki, M., Aoki, Y., Sekihara, J., Iyama, Y., Nakada, H., Iwata, T., Kichishima, T., Ebitani, T., Numada, F., Manta, H., Nakajima, H., Yamashita, T., Miyahara, K., Toyoshima, G., Yamada, K., Yamamoto, R., Ochi, S. 2020. Factors affecting red crown rot caused by Calonectria ilicicola in soybean cultivation. *Journal of General Plant Pathology* 86:363–375.

Alexander, M. *Introduction to soil microbiology*. New York: John Wiley, 1977.

Ali, B., Wang, X., Saleem, M. H., Sumaria., Hafeez, A., Afridi, M. S., Khan, S., Nisa, Z. U., Ullah, I., Amaral Júnior, A. T., Alatawi, A., Ali, S. (2022). PGPR-mediated salt tolerance in maize by modulating plant physiology, antioxidant defense, compatible solutes accumulation and bio-surfactant producing genes. *Plants* 11:345.

An, X., Liu, J., Liu, X., Ma, C., and Zhang, Q. 2022. Optimizing phosphorus application rate and the mixed inoculation of arbuscular mycorrhizal fungi and phosphate-solubilizing bacteria can improve the phosphatase activity and organic acid content in alfalfa soil. *Sustainability* 14:11342.

Apolinário, V. X. O., Dubeux, J. C. B., Lira, M. A., Ferreira, R. L. C., Mello, A. C. L., Santos, M. V. F., Sampaio, E. V. S. B., Muir, J. P. (2015). Tree legumes provide marketable wood and add nitrogen in warm-climate silvopasture systems. *Agronomy Journal* 107:1915-1921.

Araújo, A. S. F., Leite, L. F. C., Iwata, B. F., Lira Jr, M. A., Xavier, G. R., Figueiredo, M. V. B. (2012). Microbiological process in agroforestry systems. A review. *Agronomy Sustainable Development* 32: 215–226.

Araújo, V. L. V. P., Junior, M. A. L., Júnior, V. S. S., Filho, J. C. A., Fracetto, F. J. C., Dini-Andreote, F., Pereira, A. P. A., Júnior, J. P. M., Barros, F. M. R. and Fracetto, G. G. M. 2020. Bacteria from tropical semiarid temporary ponds promote maize growth under hydric stress. *Microbiological Research* 240:126564.

Backer, R., Rokem, J. S., Ilangumaran, G., Lamont, J., Praslickova, D., Ricci, E., Subramanian, S., and Smith, D. L. 2018. Plant growth-promoting rhizobacteria: context, mechanisms of action, and roadmap to commercialization of biostimulants for sustainable agriculture. *Frontiers in Plant Science* 871:1-17.

Bagayoko, M., Alvey, S., Neumann, G., Buerkert, A. 2000. Root-induced increases in soil pH and nutrient availability to field-grown cereals and legumes on acid sandy soils of Sudano-Sahelian West Africa. *Plant and Soil* 225:117-127.

Baker, K. F., Cook, R. J. *Biological control of plant pathogens*. Saint Paul: American Phytopathological Society,1974.

Barbosa, J. Z., Hungria, M., Prior, S. A., Moura, M. C., Poggere, G., and Motta, A. C. V. 2022. Improving yield and health of legume crops via co-inoculation with rhizobia and trichoderma: a global meta-analysis. *Applied Soil Ecology* 176:104493.

Barbosa, J. Z., Hungria, M., Sena, J. V. S., Poggere, G., Reis, A. R., and Corrêa, R. S. 2021. Meta-analysis reveals benefits of co-inoculation of soybean with *Azospirillum brasilense* and *Bradyrhizobium* spp. in Brazil. *Applied Soil Ecology* 163:103913.

Batista, A. M., Libardi, P. L., Alves, M. E., Prataviera, F., & Giarola, N. F. B. 2022. Electrochemical Effects on Clay Dispersion in Rhizo-and Non-rhizospheric Soils. *Journal of Soil Science and Plant Nutrition* 22:3518-3526.

Batista, A. M., Libardi, P. L., Giarola, N. F. B. 2020. Evaluation of the soil aggregation induced by the plant roots in an Oxisol by turbidimetry and water percolation. *Rhizosphere* 16:100265.

Baumert, V. L., Vasilyeva, V. A., Vladimirov, A. A., Meier, I. C., Ingrid Kögel-Knabner, I., Mueller, C. W. 2018. Root Exudates Induce Soil Macroaggregation Facilitated by Fungi in Subsoil. *Frontiers in Environmental Science* 6:140.

Bell, T., Newman, J. A., Silverman, B. W., Turner, S. L., Lilley, A. K. 2005. The contribution of species richness and composition to bacterial services. *Nature* 436:1157-1160.

Beltran-Medina, J. I., Romero-Perdomo, F., Molano-Chavez, L., Silva, A. M. M., and Estrada-Bonilla, G. A. 2022. Differential plant growth promotion under reduced phosphate rates in two genotypes of maize by a rhizobial phosphate-solubilizing strain. *Frontiers in Sustainable Food Systems* 6:1-13.

Benard, P., Kroener, E., Vontobel, P., Kaestner, A., Carminati, A. 2016. Water percolation through the root-soil interface. *Advances in Water Resources* 95:190-198.

Benard, P., Zarebanadkouki, M., Carminati, A. 2019. Physics and hydraulics of the rhizosphere network. *Journal of Plant Nutrition and Soil Science* 182:5-8.

Bender, S. F., Wagg, C., Van der Heijden, M. G. A. 2016. An underground revolution: biodiversity and soil ecological engineering for agricultural sustainability. *Trends in Ecology & Evolution* 31:440-452.

Bengough, A. G. 2012. Water dynamics of the root zone: Rhizosphere biophysics and its control on soil hydrology. *Vadose Zone Journal* 11:p.vzj2011.0111.

Berg, M., Koskella, B.2018. Nutrient- and dose-dependent microbiome-mediated protection against a plant pathogen. *Current Biology* 28:2487-2492.

Beule, L., Lehtsaar, E., Corre, M. D., Schmidt, M., Veldkamp, E., Karlovsky, P. 2020. Poplar rows in temperate agroforestry croplands promote bacteria, fungi, and denitrification genes in soils. *Frontiers in microbiology* 10:3108.

Bi, Li., Yu, Dan-Ting., Du, S., Zhang, Li-Mei., Zhang, Li-Yu., Wu, Chuan-Fa., Xiong, C., Han, Li-Li., and He, Ji-Zheng. 2021. Diversity and potential biogeochemical impacts of viruses in bulk and rhizosphere soils. *Environmental Microbiology* 23:588-599.

Bian, F., Zhong, Z., Zhang, X. Li, Q., Huang, Z. 2022. Bamboo-based agroforestry changes phytoremediation efficiency by affecting soil properties in rhizosphere and non-rhizosphere in heavy metal-polluted soil (Cd/Zn/Cu). *Journal of Soils and Sediments* 23:368-378.

Blinkov, E. A., Tsavkelova, E. A. and Selitskaya, O. V. 2014. Auxin production by the Klebsiella planticola strain TSKhA-91 and its effect on development of cucumber (Cucumis sativus L.) seeds. *Microbiology* 83:531-38.

Bravin, M. N., Garnier, C., Lenoble, V., Gérard, F., Dudal, Y., Hinsinger, P. 2012. Root-induced changes in pH and dissolved organic matter binding capacity affect copper dynamic speciation in the rhizosphere. *Geochimica et Cosmochimica Acta* 84:256-268.

Bronick, C. J., Lal, R. 2005. Soil structure and management: A review. *Geoderma* 124:3-22.

Bruns, M. A. Bacteria and archaea. In: Gentry, T. J. and Zuberer, D. A. (Eds.). *Principles and Applications of Soil Microbiology*. Elsevier Inc, 2021.

Burns, R. G., Dick, R. P. E*nzymes in the environment: activity, ecology, and applications*. Boca Raton: CRC Press, 2002.

Burr-Hersey, J. E., Ritz, K., Bengough, G. A., Mooney, S. J. 2020. Reorganisation of rhizosphere soil pore structure by wild plant species in compacted soils. *Journal of Experimental Botany* 71:6107-6115.

Cardon, Z. G., Whitbeck, J. L. *The rhizosphere: an ecological perspective*. Elsevier, 2011.

Carminati, A. 2013. Rhizosphere wettability decreases with root age: A problem or a strategy to increase water uptake of young roots? *Frontiers in Plant Science* 4:298.

Carminati, A., Moradi, A. B., Vetterlein, D., Vontobel, P., Lehmann, E., Weller, U., Vogel, H. J., Oswald, S. E. 2010. Dynamics of soil water content in the rhizosphere. *Plant and Soil* 332:163-176.

Carminati, A., Vetterlein, D. 2013. Plasticity of rhizosphere hydraulic properties as a key for efficient utilization of scarce resources. *Annals of Botany* 112:277-290.

Carminati, A., Zarebanadkouki, M., Kroener, E., Ahmed, M. A., Holz, M. 2016. Biophysical rhizosphere processes affecting root water uptake. *Annals of Botany* 118:561-571.

Carrión, V. J., Cordovez, V., Tyc, O., Etalo, D. W., Bruijn, I. D., Jager, C. L. V., Medema, M. H., Eberl, L., Raaijmakers, J. M. 2018. Involvement of Burkholderiaceae and sulfurous volatiles in disease-suppressive soils. *The ISME Journal* 12: 2307-2321.

Carrión, V. J., Perez-Jaramillo, J., Cordovez, V., Tracanna, V., De Hollander, M., Ruiz-Buck, D., Mendes, L. W., Van Ijcken, W. F., Gomez-Exposito, R., Elsayed, S. S., Mohanraju, P., Arifah, A., Van der Oost, J., Paulson, J. N., Mendes, R., Van Wezel, G. P., Medema, M. H., Raaijmakers, J. M. 2019. Pathogen-induced activation of disease-suppressive functions in the endophytic root microbiome. *Science* 366:, 606-612.

Cerri, C. E. P., Cerri, C. C., Maia, S. M. F., Cherubin, M. R., Feigl, B. J., and Lal, R. 2018. Reducing amazon deforestation through agricultural intensification in the cerrado for advancing food security and mitigating climate change. *Sustainability* 10: 1–18.

Chaiya, L., Gavinlertvatana, P., Teaumroong, N., Pathom-aree, W., Chaiyasen, A., Sungthong, R., Lumyong, S. 2021. Enhancing Teak (Tectona grandis) Seedling Growth by Rhizosphere Microbes: A Sustainable Way to Optimize Agroforestry. *Microorganisms* 9: 1990.

Channarayappa, C., & Biradar, D. P. *Soil basics, management, and rhizosphere engineering for sustainable agriculture*. CRC Press, 2018.

Chen, C., Chen, H. Y. H., Chen, X., Huang, Z. 2019a. Meta-analysis shows positive effects of plant diversity on microbial biomass and respiration. *Nature Communications* 10: 1332.

Chen, C., Wu, L., Harbottle, M. 2019b. Influence of biopolymer gel-coated fibres on sand reinforcement as a model of plant root behaviour. *Plant and Soil* 438:361-375.

Chen, Xue-Ping., Zhu, Yong-Guan., Xia, Y., Shen, Ju-Pei., and He, Ji-Zheng. 2008. Ammonia-oxidizing archaea: important players in paddy rhizosphere soil? *Environmental Microbiology* 10: 1978–87.

Cheng, W. and Gershenson, A. Carbon Fluxes in the Rhizosphere. In: Cardon, Z. and Whitbeck, J. (Eds.). *The Rhizosphere: An Ecological Perspective*. Elsevier Inc, 2007.

Cong, W., Hoffland, E., Li, L., Janssen, B., Van der Werf, W. 2015. Intercropping affects the rate of decomposition of soil organic matter and root litter. *Plant and Soil* 39: 399-411.

Cox, C. J., Foster, P. G., Hirt, R. P., Harris, S. R., and Embley, T. M. 2008. The Archaebacterial Origin of Eukaryotes. *Proceedings of the National Academy of Sciences of the United States of America* 105: 20356–20361.

Cruz, B. C., Furrer, J. M., Guo, Y. S., Dougherty, D., Hinestroza, H. F., Hernandez, J. S., Gage, D. J., Cho, Y. K., Shor, L. M. 2017. Pore-scale water dynamics during drying and the impacts of structure and surface wettability. *Journal of the American Water Resources Association* 53: 5585-5600.

Czarnes, S., Dexter, A. R., Bartoli, F. 2000. Wetting and drying cycles in the maize rhizosphere under controlled conditions. Mechanics of the root-adhering soil. *Plant and Soil* 221:253-271.

Daly, K. R., Mooney, S. J., Bennett, M. J., Crout, N. M. J., Roose, T., Tracy, S. R. 2015. Assessing the influence of the rhizosphere on soil hydraulic properties using X-ray computed tomography and numerical modelling. *Journal of Experimental Botany,* 66:2305-2314.

De Stefano, A., Jacobson, M. G. 2018. Soil carbon sequestration in agroforestry systems: a meta-analysis. *Agroforestry Systems* 92:285-299.

Decraemer, W. and Hunt, D. J. Structure and classification. In: Perry, R. N. and Moens, M., (Eds.). *Plant Nematology*. Wallingford, Oxfordshire: CAB International, 2006.

Demenois, J., Carriconde, F., Bonaventure, P., Maeght, J. L., Stokes, A., Rey, F. 2018. Impact of plant root functional traits and associated mycorrhizas on the aggregate stability of a tropical Ferralsol. *Geoderma* 312:6-16.

Deng, P., Fan, W., Wang, H., Ti, J., Xu, X. 2022. Chinese Torreya agroforestry alters the rhizosphere soil bacterial communities and potential functions. *Applied Soil Ecology* 177:104504.

Dhaouadi, S., Rouissi, W., Mougou-Hamdane, A., Nasraoui, B. 2018. Evaluation of biocontrol potential of Achromobacter xylosoxidans against Fusarium wilt of melon. *European Journal of Plant Pathology* 154: 179–88.

Dimkić, I., Janakiev, T., Petrović, M., Degrassi, G., Fira, D. 2022. Plant-associated Bacillus and Pseudomonas antimicrobial activities in plant disease suppression via biological control mechanisms—A review. *Physiological and Molecular Plant Pathollogy* 117:101754.

Dollinger, J., Jose, S. 2018. Agroforestry for Soil Health. *Agroforestry Systems* 92: 213–219.

Drogue, B., Combes-Meynet, E., Moënne-Loccoz, Y., Wisniewski-Dyé, F., Prigent-Combaret, C. Control of the cooperation between plant growth-promoting rhizobacteria and crops by rhizosphere signals. In. Bruijn, F. J. (Ed.). *Molecular microbial ecology of the rhizosphere.* New Jersey: John Wiley & Sons, 2013.

Duchene, O., Vian, J. F., Celette, F. 2017. Intercropping with legume for agroecological cropping systems: complementarity and facilitation processes and the importance of soil microorganisms. A review. *Agriculture Ecosystems & Environment* 240:148-161.

Durán, P., G. Tortella, S., Viscardi, P. J., Barra, V. J., Carrión, La. Mora, Pozo, M. J. 2018. Microbial community composition in take-all suppressive soils. *Frontiers in Microbiology* 9:2198.

Ehrmann, J., Ritz, K. 2014. Plant: soil interactions in temperate multi-cropping production systems. *Plant and Soil* 376:1-29.

Eisenhauer, N., Lanoue, A., Strecker, T., Scheu, S., Steinauer, K., Thakur, M. P., Mommer, L. 2017. Root biomass and exudates link plant diversity with soil bacterial and fungal biomass. *Scientific Reports* 7:44641.

Elevitch, C. R., Mazaroli, D. N., Ragone, D. 2018. Agroforestry Standards for Regenerative Agriculture. *Sustainability* 10:3337.

El-Ramady, H., Brevik, E. C., Elbasiouny, H., Elbehiry, F., El-Henawy, A., Faizy, S. E. D., Elsakhawy, T., Omara, A. E. D., Amer, M., and Eid, Y. 2021. Soils, biofortification, and human health under Covid-19: challenges and opportunities. *Frontiers in Soil Science* 1:1-9.

Eme, L., Spang, A., Lombard, J., Stairs, C. W., Ettema, T. J. G. 2017. Archaea and the origin of Eukaryotes. *Nature Reviews Microbiology* 15:711-723.

Erel, R., Bérard, A., Capowiez, L., Doussan, C., Arnal, D., Souche, G., Gavaland, A., Fritz, C., Visser, E. J. W., Salvi, S., Le Marié, C., Hund, A., Hinsinger, P. 2017. Soil type determines how root and rhizosphere traits relate to phosphorus acquisition in field-grown maize genotypes. *Plant and Soil* 412:115-132.

Escalas, A., Hale, L., Voordeckers, J. W., Yang, Y., Firestone, M. K., Alvarez-Cohen, L., and Zhou, J. 2019. Microbial functional diversity: from concepts to applications. *Ecology and Evolution* 9:12000-12016.

Estrada-Bonilla, G. A., Durrer, A., and Cardoso, E. J. B. N. 2021. Use of compost and phosphate-solubilizing bacteria affect sugarcane mineral nutrition, phosphorus availability, and the soil bacterial community. *Applied Soil Ecology* 157:103760.

Expósito, R. G., de Bruijn, I., Postma, J., Raaijmakers, J. M. 2017. Current insights into the role of rhizosphere bacteria in disease suppressive soils. *Frontiers in Microbiology* 8:2529.

Fageria, N. K., Stone, L. F. 2006. Physical, Chemical, and Biological Changes in the Rhizosphere and Nutrient Availability. *Journal of Plant Nutrition* 29:1327-1356.

FAO. 2021. *New standards to curb the global spread of plant pests and diseases*, January 22, 2023, https://www.fao.org/news/story/en/item/1187738/.

Feeney, D. S., Crawford, J. W., Daniell, T., Hallett, P. D., Nunan, N., Ritz, K., Rivers, M., Young, I. M. 2006. Three-dimensional microorganization of the soil-root-microbe system. *Microbial Ecology* 52:151-158.

Filo, A., Sabbatini, P., Sundin, G. W., Zabadal, T. J., Safir, G. R., Cousins, P. S. 2013. Grapevine crown gall suppression using biological and genetic engineering: a review of recent research. *American Journal of Enology and Viticulture* 64:1–14.

Forterre, P., and Prangishvili, D. 2009. The origin of viruses. *Research in Microbiology* 160: 466–72.

Fortuna, A. *The soil biota*. Nature Education Knowledge, 2012.

Francl, L. J. *The disease triangle: a plant pathological paradigm revisited*. The Plant Health Instructor, 2001.

Frantzeskakis, L., Di Pietro, A., Rep, M., Schirawski, J., Wu, C.-H., Panstruga, R. 2020. Rapid evolution in plant-microbe interactions – a molecular genomics perspective. *New Phytologist* 225:1134-1142.

Gama-Rodrigues, A. C., Müller, M. W., Gama-Rodrigues, E. F., Mendes, F. A. T. 2021. Cacao-based agroforestry systems in the Atlantic Forest and Amazon Biomes: an ecoregional analysis of land use. *Agricultural Systems* 194:103270.

Gebremikael, M. T., Steel, H., Buchan, D., Bert, W., and Neve, S. D. 2016. Nematodes enhance plant growth and nutrient uptake under C and N-rich conditions. *Scientific Reports* 6: 1–10.

George, T. S., Gregory, P. J., Wood, M., Read, D., Buresh, R. J. 2002. Phosphatase activity and organic acids in the rhizosphere of potential agroforestry species and maize. *Soil Biology and Biochemistry* 34:1487-1494.

Ghezzehei, T. A., Albalasmeh, A. A. 2015. Spatial distribution of rhizodeposits provides built-in water potential gradient in the rhizosphere. *Ecological Modelling* 298:53-63.

Giles, C. D., Brown, L. K., Adu, M. O., Mezeli, M. M., Sandral, G. A., Simpson, R. J., Wendler, R., Shand, C. A., Menezes-Blackburn, D., Darch, T., Stutter, M. I., Lumsdon, D. G., Zhang, H., Blackwell, M. S. A., Wearing, C., Cooper, P., Haygarth, P. M., George, T. S. 2017. Response-based selection of barley cultivars and legume species for complementarity: Root morphology and exudation in relation to nutrient source. *Plant Science* 255:12-28.

Glick, B. R. 2012. Plant growth-promoting bacteria: mechanisms and applications. *Scientifica* 2012: 963401.

Graham, J. H., Strauss, S. L. Biological control of soilborne plant pathogens and nematodes James. In: Gentry, T. J., Fuhrmann, J. J., Zuberer, D. A. *Principles and Applications of Soil Microbiology*. Elsevier, 2021.

Gregory, P. J., Hinsinger, P. 1999. New approaches to studying chemical and physical changes in the rhizosphere: an overview. *Plant and Soil* 211: 1-9.

Gregory, P. J., George, T. S., and Paterson, E. 2022. New methods for new questions about rhizosphere/plant root interactions. *Plant and Soil* 476:699-712.

Gu, Y., Banerjee, S., Dini-Andreote, F., Xu, Y., Shen, Q., Jousset, A., Wei, Z. 2022. Small changes in rhizosphere microbiome composition predict disease outcomes earlier than pathogen density variations. *The ISME Journal* 16:2448-2456.

Guo, L., Yu, Z., Li, Y., Xie, Z., Wang, G., Liu, X., Liu, J., Liu, J., Jin, J. 2022. Plant phosphorus acquisition links to phosphorus transformation in the rhizospheres of soybean and rice grown under CO_2 and temperature co-elevation. *Science of The Total Environment* 823:153558.

Hao, C., Dungait, J. A. J., Wei, X., Ge, T., Kuzyakov, Y., Cui, Z., Tian, J., Zhang, F. 2022. Maize root exudate composition alters rhizosphere bacterial community to control hotspots of hydrolase activity in response to nitrogen supply. *Soil Biology and Biochemistry* 170: 108717.

Hartmann, A., Rothballer, M., and Schmid, M. 2008. Lorenz Hiltner, a pioneer in rhizosphere microbial ecology and soil bacteriology research. *Plant and Soil* 312:7-14.

Hauggaard-Nielsen, H., Ambus, P., Jensen, E. 2003. The comparison of nitrogen use and leaching in sole cropped versus intercropped pea and barley. *Nutrient Cycling in Agroecosystems* 65:289-300.

Helliwell, J. R., Sturrock, C. J., Mairhofer, S., Craigon, J., Ashton, R. W., Miller, A. J., Whalley, W. R., Mooney, S. J. 2017. The emergent rhizosphere: Imaging the development of the porous architecture at the root-soil interface. *Scientific Reports* 7:14875.

Hernández-Morcillo, M., Burgess, P., Mirck, J., Pantera, A., Plieninger, T. 2018. Scanning agroforestry-based solutions for climate change mitigation and adaptation in Europe. *Environmental Science & Policy* 80:44–52.

Hiltner, L. 1904. Über neuere Erfahrungen und Probleme auf dem Gebiete der Bodenbakteriologie unter besonderer Berücksichtigung der Gründüngung und Brache [On recent experiences and problems in the field of soil bacteriology with special reference to green manure and fallow land]. *Arbeiten Der Deutschen Landwirtschaftlichen Gesellschaft [Publications of the German Agricultural Society]*, 98:59-78.

Hinsinger, P., Bengough, A. G., Vetterlein, D., and Young, I. M. 2009. Rhizosphere: biophysics, biogeochemistry and ecological relevance. *Plant and Soil* 321:117-52.

Hopkins, R., Wojdak, J. M., Belden, L. K. 2017. Defensive symbionts mediate host-parasite interactions at multiple scales. *Trends in Parasitology* 33:53-64.

Hosseini, F., Mosaddeghi, M. R., Hajabbasi, M. A., Sabzalian, M. R. 2015. Influence of tall fescue endophyte infection on structural stability as quantified by high energy moisture characteristic in a range of soils. *Geoderma* 249:87-99.

Hu, J., Wei, Z., Friman, V. P., Gu, S. H., Wang, X. F., Eisenhauer, N., Yang, T-J., Ma, J., Shen, Q. R., Xu, Y. C., Jousset, A. 2016. Probiotic diversity enhances rhizosphere microbiome function and plant disease suppression. *American Society for Microbiology* 7:e01790-16.

Hu, L., Robert, C. A. M., Cadot, S., Zhang, X., Ye, M., Li, B., Manzo, D., Cherver, N., Steinger, T., van der Heijden, M. G. A., Schlaeppi, K., Erb, M. 2018. Root exudate metabolites drive plant-soil feedbacks on growth and defense by shaping the rhizosphere microbiota. *Nature Communications* 9:2738.

Huang, K., Jiang, Q., Liu, L., Zhang, S., Liu, C., Chen, H., Ding, W., and Zhang, Y. 2020. Exploring the key microbial changes in the rhizosphere that affect the occurrence of tobacco root-knot nematodes. *AMB Express* 10:1-11.

Huang, X-F., Chaparro, J. M., Reardon, K. F., Zhang, R., Shen, Q., Vivanco, J. M. 2014. Rhizosphere interactions: root exudates, microbes, and microbial communities. *Botany* 92:267-275.

Hungria, M., Campo, R. J., Souza, E. M. and Pedrosa, F. O. 2010. Inoculation with selected strains of *Azospirillum brasilense* and *A. lipoferum* improves yields of maize and wheat in Brazil. *Plant and Soil* 331:413-25.

Isaac, M. E., Borden, K. A. 2019. Nutrient acquisition strategies in agroforestry systems. *Plant and Soil* 444:1-19.

Janczarek, M., Rachwał, K., Marzec, A., Grządziel, J., Palusińska-Szysz M. 2015. Signal molecules and cell-surface components involved in early stages of the legume–rhizobium interactions. *Applied Soil Ecology* 85:94-113.

Jiang, F., Zhang, L., Zhou, J., George, T. S. and Feng, G. 2021. Arbuscular mycorrhizal fungi enhance mineralisation of organic phosphorus by carrying bacteria along their extraradical hyphae. *New Phytologist* 230:304-315.

Jones, D. L., and Hinsinger, P. 2008. The rhizosphere: complex by design. *Plant and Soil* 312:116.

Jones, D. L., Hodge, A., Kuzyakov, Y. 2004. Plant and mycorrhizal regulation of rhizodeposition. *New Phytologist* 163:459-80.

Jones, D. L., Nguyen, C., and Finlay, R. D. 2009. Carbon flow in the rhizosphere: carbon trading at the soil-root interface. *Plant and Soil* 321:5-33.

Jones, J. T., Haegeman, A., Danchin, E. G. J., Gaur, H. S., Helder, J., Jones, M. G. K., Kikuchi, T., Manzanilla-López, R., Palomares-Rius, J. E., Wesemael, W. M. L., Perry, R. N. 2013. Top 10 plant-parasitic nematodes in molecular plant pathology. *Molecular Plant Pathology* 14:946-61.

Jongen, M., Albadran, B., Beyschlag, W., and Unger, S. 2022. Can arbuscular mycorrhizal fungi mitigate drought stress in annual pasture legumes? *Plant and Soil* 472:295-310.

Jung, J., Kim, Jun-Seob., Taffner, J., Berg, G., and Ryu, Choong-Min. 2020. Archaea, tiny helpers of land plants. *Computational and Structural Biotechnology Journal* 18:2494-2500.

Kristensen, N. B., Bryrup, T., Allin, K. H., Nielsen, T., Hansen, T. H., Pedersen, O. 2016. Alterations in fecal microbiota composition by probiotic supplementation in healthy adults: a systematic review of randomized controlled trials. *Genome Medicine* 8:52.

Kroener, E., Holz, M., Zarebanadkouki, M., Ahmed, M. A., Carminati, A. 2018. Effects of mucilage on rhizosphere hydraulic functions depend on soil particle size. *Vadose Zone Journal* 17:170056.

Kroener, E., Zarebanadkouki, M., Bittelli, M., Carminati, A. 2016. Simulation of root water uptake under consideration of nonequilibrium dynamics in the rhizosphere. *Water Resources Research*. 52:5755-5770.

Kroener, E., Zarebanadkouki, M., Kaestner, A., Carminati, A. 2014. Nonequilibrium water dynamics in the rhizosphere: How mucilage affects water flow in soils. *Water Resources Research* .50:6479-6495.

Kumar, A. J. P., Verma, J. P. *The Role of Microbes to Improve Crop Productivity and Soil Health BT* - Ecological Wisdom Inspired Restoration Engineering. Singapore: Springer, 2019.

Kumar, A., Dorodnikov, M., Splettstößer, T., Kuzyakov, Y., Pausch, J. 2017. Effects of maize roots on aggregate stability and enzyme activities in soil. *Geoderma* 306:50-57.

Kumar, M., Mondal, R., Gurusamy, S., Balakrishnan, S., Natarajan. Plant-microbiome interaction and the effects of biotic and abiotic components in agroecosystem. In: Singh, D., Gupta, V., Prabha, R. (Eds.). *Microbial interventions in agriculture and environment*. Singapore: Springer, 2019.

Kuzyakov, Y., Razavi, B. S. 2019. Rhizosphere size and shape: Temporal dynamics and spatial stationarity. *Soil Biology and Biochemistry* 135:343-360.

Kwak, M. J., Kong, H. G., Choi, K., Kwon, S. K., Song, J. Y., Lee, J., Lee, P. A., Choi, S. Y., Seo, M., Lee, H. J., Jung, E. J., Park, H., Roy, N., Kim, H., Lee, M. M., Rubin, E. M., Lee, S. W., Kim, J. F. 2018. Rhizosphere microbiome structure alters to enable wilt resistance in tomato. *Nature Biotechnology* 36:1117.

Lebeis, S. L., Paredes, S. H., Lundberg, D. S., Breakfield, N., Gehring, J., McDonald, M., Malfatti, S., Del Rio, T. G., Jones, C. D., Tringe, S. G., Dangl, J. L. 2015. Salicylic acid modulates colonization of the root microbiome by specific bacterial taxa. *Science* 349:860-864.

Lee, S. M., Kong, H. G., Song, G. C., Ryu, C. M. 2021. Disruption of Firmicutes and Actinobacteria abundance in tomato rhizosphere causes the incidence of bacterial wilt disease. *The ISME Journal* 15:330-347.

Lehmann, A., Leifheit, E. F., Rillig, M. C. Mycorrhizas and soil aggregation. In: Johnson, N.; Gehring, C.; Jansa, J. *Mycorrhizal mediation of soil: Fertility, structure, and carbon storage*. Elsevier, 2017.

Li, X., Yao, S., Bian, Y., Jiang, X., Song, Y. 2020. The combination of biochar and plant roots improves soil bacterial adaptation to PAH stress: Insights from soil enzymes, microbiome, and metabolome. *Journal of Hazardous Materials* 400:123227.

Li, B., Zhang, Q., Chen, Y., Su, Y., Sun, S., Chen, G. 2021. Different crop rotation systems change the rhizosphere bacterial community structure of *Astragalus membranaceus* (Fisch) Bge. var. mongholicus (Bge.) Hsiao. *Applied Soil Ecology* 166:104003.

Liang, S., Deng, J., Jiang, Y., Wu, S., Zhou, Y., Zhu, W. 2020. Functional distribution of bacterial community under different land use patterns based on FaProTax function prediction. *Polish Journal of Environmental Studies* 29: 1245-1261.

Lira Junior, M. A., Fracetto, F. J. C., Ferreira, J. S., Silva, M. B., Fracetto, G. G. M. 2020. Legume-based silvopastoral systems drive C and N soil stocks in a subhumid tropical environment. *Catena* 189:104508.

Liu, C., Jin, Y., Hu, Y., Tang, J., Xiong, Q., Xu, M., Bibi, F., Beng, K. C. 2019. Drivers of soil bacterial community structure and diversity in tropical agroforestry systems. *Agriculture, Ecosystems & Environment* 278: 24-34.

Liu, C., Wang, Q. W., Jin, Y., Tang, J., Lin, F., Olatunji, O. A. 2021. Perennial cover crop biomass contributes to regulating soil P availability more than rhizosphere P-mobilizing capacity in rubber-based agroforestry systems. *Geoderma* 401:115218.

Liu, X., Zhang, S., Jiang, Q., Bai, Y., Shen, G., Li, S., Ding, W. 2016. Using community analysis to explore bacterial indicators for disease suppression of tobacco bacterial wilt. *Scientific Reports* 6:36773.

Liu, S., Yu, J., Yasufuku, N. 2019. Physically based soil water characteristic curves function for soils with inner porosity. *Archives of Agronomy and Soil Science* 65:537-548.

Liu, H., Hao, Z., Yuan, Y., Li, C., Zhang, J. 2022a. Application of mineral phosphorus fertilizer influences rhizosphere chemical and biological characteristics. *Archives of Agronomy and Soil Science* 1-14.

Liu, S., He, F., Kuzyakov, Y., Xiao, H., Hoang, D. T. T., Pu, S., Razavi, B. S. 2022b. Nutrients in the rhizosphere: A meta-analysis of content, availability, and influencing factors. *Science of The Total Environment* 826:153908.

Loper, J. E., Hassan, K. A., Mavrodi, D. V., Davis II, E. W., Lim, C. K., Shaffer, B. T., Elbourne, L. D., Stockwell, V. O., Hartney, S. L., Breakwell, K., Henkels, M. D., Tetu, S. G., Rangel, L. I., Kidarsa, T. A., Wilson, N. L., Van de Mortel, J. E., Song, C., Blumhagen, R., Radune, D., Hostetler, J. B., Brinkac, L. M., Durkin, A. S., Kluepfel, D. A., Wechter, W. P., Anderson, A. J., Kim, Y. C., Pierson III, L. S., Pierson, E. A.,

Lindow, S. E., Kobayashi, D. Y., Raaijmakers, J. M., Weller, D. M., Thomashow, L. S., Allen, A. E., Paulsen, I. T. 2012. Comparative genomics of plant-associated *Pseudomonas* spp.: insights into diversity and inheritance of traits involved in multitrophic interactions. *PLoS Genetics* 8:e1002784.

Lopes, C. M., Silva, A. M. M., Estrada-Bonilla, G. A., Ferraz-Almeida, R., Vieira, J. L. V., Otto, R., Vitti, G. C., Cardoso, E. J. B. N. 2021. Improving the fertilizer value of sugarcane wastes through phosphate rock amendment and phosphate-solubilizing bacteria inoculation. *Journal of Cleaner Production* 298: 126821.

Lovell, S. T., Dupraz, C., Gold, M., Jose, S., Revord, R., Stanek, E., Wolz K. J. 2018. Temperate agroforestry research: considering multifunctional woody polycultures and the design of long-term field trials. *Agroforest Systems* 92:1397-1415.

Mathesius, U., Costa, S. R. 2021. Plant signals differentially affect rhizosphere nematode populations. *Journal of Experimental Botany* 72:3496-3499.

Mawarda, P. C., Roux, X. L., van Elsas, J. D., Salles, J. F. 2020. Deliberate introduction of invisible invaders: a critical appraisal of the impact of microbial inoculants on soil microbial communities. *Soil Biology and Biochemistry* 148:107874.

Maxton, A., Singh, P., and Masih, S. A. 2018. ACC deaminase-producing bacteria mediated drought and salt tolerance in *Capsicum annuum*. *Journal of Plant Nutrition* 41:574-583.

Mazurier, S., Corberand, T., Lemanceau, P., and Raaijmakers, J. M. 2009. Phenazine antibiotics produced by fluorescent pseudomonads contribute to natural soil suppressiveness to fusarium wilt. *ISME Journal* 3:977-91.

McNear, D. H. 2013. The Rhizosphere - Roots, soil and everything in between meeting the global challenge of sustainable food, fuel and fiber production. *Nature Education Knowledge* 4:1-15.

Meharg, A. A., Killham, K. 1988. A comparison of carbon flow from pre-labeled and pulse-labeled plants. *Plant and Soil* 112:225-231.

Mendes, I. C., Chaer, G. M., Reis-Junior, F. B., Sousa, D. M. G., Silva, O. D. D., Oliveira, M. I., Malaquias, J. V. *Tecnologia BioAS: uma maneira simples e eficiente de avaliar a saúde do solo [BioAS Technology: a simple and efficient way to assess soil health].* Planaltina: Embrapa Cerrados, 2021.

Mendes, L. W., Raaijmakers, J. M., De Hollander, M., Mendes, R., Tsai, S. M. 2018. Influence of resistance breeding in common bean on rhizosphere microbiome composition and function. *The ISME Journal* 12:212-224.

Mendes, R., Garbeva, P., and Raaijmakers, J. M. 2013. The rhizosphere microbiome: significance of plant beneficial, plant pathogenic, and human pathogenic microorganisms. *FEMS Microbiology Reviews* 37:634-63.

Mendes, R., Kruijt, M., De Bruijn, I., Dekkers, E., Van Der Voort, M., Schneider, J. H. M., Piceno, Y. M., Desantis, T. Z., Andersen, G. L., Bakker, P. A. H. M. Raaijmakers J. M. 2011. Deciphering the rhizosphere microbiome for disease suppressive bacteria. *Science* 332:1097-1100.

Mendes, R., Raaijmakers, J. M. 2015. Cross-kingdom similarities in microbiome functions. *The ISME Journal* 9:1905-1907.

Mercado-Blanco, J., Abrantes, I., Barra Caracciolo, A., Bevivino, A., Ciancio, A., Grenni, P., Hrynkiewicz, K., Kredics, L. Proença, D. N. 2018. Belowground Microbiota and the Health of Tree Crops. *Frontiers of Microbiololy* 9:1006.

Merino-Martín, L., Stokes, A., Gweon, H. S., Moragues-Saitua, L., Staunton, S., Plassard, C., Oliver, A., Le Bissonnais, Y., Griffiths, R. I. 2021. Interacting effects of land use type, microbes and plant traits on soil aggregate stability. *Soil Biology and Biochemistry* 154:108072.

Mishra, J., Singh, R., and Arora, N. K. 2017. Alleviation of heavy metal stress in plants and remediation of soil by rhizosphere microorganisms. *Frontiers in Microbiology* 8:1-7.

Mohanty, P., Singh, P. K., Chakraborty, D., Mishra, S., and Pattnaik, R. 2021. Insight into the role of PGPR in sustainable agriculture and environment. *Frontiers in Sustainable Food Systems* 5:1-12.

Moissl-Eichinger, C., Pausan, M., Taffner, J., Berg, G., Bang, C., and Schmitz, R. A. 2018. Archaea are interactive components of complex microbiomes. *Trends in Microbiology* 26:70-85.

Moreno-Galván, A. E., Cortés-Patiño, S., Romero-Perdomo, F., Uribe-Vélez, D., Bashan, Y., and Bonilla, R. B. 2020. Proline accumulation and glutathione reductase activity induced by drought-tolerant rhizobacteria as potential mechanisms to alleviate drought stress in guinea grass. *Applied Soil Ecology* 147:103367.

Mortimer, R., Saj, S., David, C. 2018. Supporting and regulating ecosystem services in cacao agroforestry systems. *Agroforestry Systems* 92:1639-1657.

Moura, E. G., Aguiar, A. C. F., Piedade, A. R., Rousseau, G. X. 2015. Contribution of legume tree residues and macrofauna to the improvement of abiotic soil properties in the eastern Amazon. *Applied Soil Ecology* 86:91-99.

Mukhtar, T., Afridi, M. S., Mc Arthur, R., Van Hamme, J. D., Rineau, F., Mahmood, T., Aman., Sumaria., Zahid, M., Salam, A., Khan, M. N., Ali, F., Mehmood, S., Bangash, N., Chaudhary, H. J. 2018. Draft genome sequence of Bacillus pumilus SCAL1, an endophytic heat-tolerant plant growth-promoting bacterium. *Genome Announcements* 6:e00306-e00318.

Munroe, J. W., Soto, G., Virginio Filho, E. D. M., Fulthorpe, R., Isaac, M. E. 2015. Soil microbial and nutrient properties in the rhizosphere of coffee under agroforestry management. *Applied Soil Ecology* 93:40-46.

Nath, A. J., Lal, R., Das, A. K. 2015. Ethnopedology and soil properties in bamboo (*Bambusa* sp.) based agroforestry system in North East India. *Catena* 135:92-99.

Naveed, M., Brown, L. K., Raffan, A. C., George, T. S., Bengough, A. G., Roose, T., Sinclair, I., Koebernick, N., Cooper, L., Hackett, C. A., Hallett, P. D. 2017. Plant exudates may stabilize or weaken soil depending on species, origin and time. *European Journal of Soil Science* 68:806-816.

Nazari, M. 2021. Plant mucilage components and their functions in the rhizosphere. *Rhizosphere* 18:100344.

Neumann, G., Römheld, V. Rhizosphere chemistry in relation to plant nutrition. In: *Marschner's mineral nutrition of higher plants*. Academic Press, 2012.

Neumann, A., Schmidtke, K., Rauber, R. 2007. Effects of crop density and tillage system on grain yield and N uptake from soil and atmosphere of sole and intercropped pea and oat. *Field Crops Research* 100:285-293.

Niu, B., Wang, W., Yuan, Z., Sederoff, R. R., Sederoff, H., Chiang, V. L., Borriss, R. 2020. Microbial interactions within multiple-strain biological control agents impact soil-borne plant disease. *Frontiers in Microbiology* 11: 585404.

Nouioui, I., Cortés-Albayay, C., Carro, L., Castro, J. F., Gtari, M., Ghodhbane-Gtari, F., Klenk, H. P., Tisa, L. S., Sangal, V., and Goodfellow, M. 2019. Genomic insights into plant-growth-promoting potentialities of the genus Frankia. *Frontiers in Microbiology* 10:1-15.

Oburger, E., Schmidt, H. 2016. New methods to unravel rhizosphere processes. *Trends in Plant Science* 21:243-55.

Oleghe, E., Naveed, M., Baggs, E. M., Hallett, P. D. 2017. Plant exudates improve the mechanical conditions for root penetration through compacted soils. *Plant and Soil* 421:19-30.

Pardo-Díaz, S., Romero-Perdomo, F., Mendoza-Labrador, J., Delgadillo-Duran, D., Castro-Rincon, E., Silva, A. M. M., Rojas-Tapias, D. F., Cardoso, E. J. B. N., Estrada-Bonilla, G. A. 2021. Endophytic PGPB improves plant growth and quality, and modulates the bacterial community of an intercropping system. *Frontiers in Sustainable Food Systems* 5:1-10.

Paterson, E. 2003. Importance of rhizodeposition in the coupling of plant and microbial productivity. *European Journal of Soil Science* 54:741-750.

Paterson, E., Gebbing, T., Abel, C., Sim, A., and Telfer, G. 2007. Rhizodeposition shapes rhizosphere microbial community structure in organic soil. *New Phytologist* 173:600-610.

Pérez-Jaramillo, J. E., Carrión, V. J., de Hollander, M., Raaijmakers, J. M. 2018. The wild side of plant microbiomes. *Microbiome* 6:1-6.

Philippot, L., Raaijmakers, J. M., Lemanceau, P., and. Van Der Putten, W. H. 2013. Going back to the roots: the microbial ecology of the rhizosphere. *Nature Reviews Microbiology* 11:789-799.

Prajapati, S., Kumar, N., Kumar, S., Lakharan, L., Maurya, S. 2020. Biological control a sustainable approach for plant diseases management: A review. *Journal of Pharmacognosy and Phytochemistry* 9:1514-1523.

Pratama, A. A., Terpstra, J., Oliveria, A. L. M., and Salles, J. F. 2020. The role of rhizosphere bacteriophages in plant health. *Trends in Microbiology* 28:709-718.

Prescott, C. E., Grayston, S. J. 2013. Tree species influence on microbial communities in litter and soil: current knowledge and research needs. *Forest Ecology and Management* 309:19-27.

Pudjiasmanto, B., Purnomo, D., Cahyani, V. R. 2018. Diversity of arbuscular mycorrhiza and maize yield in Cajeput agroforestry system with different fertilizer management. *Bulgarian Journal of Agricultural Science* 24:611-616.

Raaijmakers, J. M., Mazzola, M. 2016. Soil immune responses. *Science* 352:1392-1393.

Rehman, Z. U., and Leiknes, T. O. 2018. Quorum-quenching bacteria isolated from red sea sediments reduce biofilm formation by Pseudomonas aeruginosa. *Frontiers in Microbiology* 9:1-13.

Richter, D. B., Oh, N-H., Fimmen, R., Jackson, J. The rhizosphere and soil formation. In: *The rhizosphere.* Academic Press, 2007.

Robertson-Albertyn, S., Terrazas R. A., Balbirnie, K., Blank, M., Janiak, A., Szarejko, I., Chmielewska, B., Karcz, J., Morris, J., Hedley, P. E., George, T. S., Bulgarelli, D. 2017. Root hair mutations displace the barley rhizosphere microbiota. *Frontiers in Plant Science* 8:1-15.

Rodrigues, M., Withers, P. J. A., Soltangheisi, A., Vargas, V., Holzschuh, M., Pavinato, P. S. 2021. Tillage systems and cover crops affecting soil phosphorus bioavailability in Brazilian Cerrado Oxisols. *Soil and Tillage Research* 205:104770.

Romero-Perdomo, F., Beltrán, I., Mendoza-Labrador, J., Estrada-Bonilla, G., and Bonilla, R. 2021. Phosphorus nutrition and growth of cotton plants inoculated with growth-promoting bacteria under low phosphate availability. *Frontiers in Sustainable Food Systems* 4:1-11.

Saleem, M., Hu, J., Jousset, A. 2019. More than the sum of its parts: microbiome biodiversity as a driver of plant growth and soil health. *Annual Review of Ecology, Evolution, and Systematics* 50:145-168.

Santos, M. S., Nogueira, M. A., and Hungria, M. 2019. Microbial inoculants: reviewing the past, discussing the present and previewing an outstanding future for the use of beneficial bacteria in agriculture. *AMB Express* 9:1-22.

Santos-Torres, M., Romero-Perdomo, F., Mendoza-Labrador, J., Gutiérrez, A. Y., Vargas, C., Castro-Rincon, E., Caro-Quintero, A., Uribe-Velez, D., and Estrada-Bonilla, G. A. 2021. Genomic and phenotypic analysis of rock phosphate-solubilizing rhizobacteria. *Rhizosphere* 17:100290.

Savary, S., Willocquet, L., Pethybridge, S. J., Esker, P., McRoberts, N., Nelson, A. 2019. The global burden of pathogens and pests on major food crops. *Nature Ecology & Evolution* 3:430-439.

Schlatter, D., Kinkel, L., Thomashow, L., Weller, D., Paulitz, T. 2017. Disease suppressive soils: new insights from the soil microbiome. *Phytopathology* 107:1284-1297.

Schmidt, J. E., DuVal, A., Isaac, M. E. Hohmann, P.2022. At the roots of chocolate: understanding and optimizing the cacao root-associated microbiome for ecosystem services. A review. *Agronomy for Sustainable Development* 42:14.

Sender, R., Fuchs, S., and Milo, R. 2016. Revised estimates for the number of human and bacteria cells in the body. *Plos Biology* 14:1-14.

Shen, J., Yuan, L., Zhang, J., Li, H., Bai, Z., Che, X., Zhang, W., Zhang, F. 2011. Phosphorus Dynamics: From Soil to Plant. *Plant Physiology* 156: 997-1005.

Silva, A. M. M., Estrada-Bonilla, G. A., Lopes, C. M., Matteoli, F. P., Cotta, S. R., Feiler, H. P., Rodrigues, Y. F., and Cardoso, E. J. B. N. 2022. Does organomineral fertilizer combined with phosphate-solubilizing bacteria in sugarcane modulate soil microbial community and functions? *Microbial Ecology* 84:539-555.

Silva, A. M. M., Pimenta, L. S., Qi, X., and Cardoso, E. J. B. N. 2023. Economic gains using organic P source and inoculation with P-solubilizing bacteria in sugarcane. *Brazilian Journal of Agricultural and Environmental Engineering* 27:101-107.

Singh, B. K., Millard, P., Whiteley, A. S., and Murrell, J. C. 2004. Unravelling rhizosphere-microbial interactions: opportunities and limitations. *Trends in Microbiology* 12:386-393.

Singh, P. P., Shin, Y. C., Park, C. S., and Chung, Y. R. 1999. Biological control of fusarium wilt of cucumber by chitinolytic bacteria. *Phytopathology* 89:92-99.

Siqueira, J. O., Souza, F. A., Cardoso, E. J. B. N., Tsai, S M. *Micorrizas: 30 anos de pesquisas no Brasil [Mycorrhizas: 30 years of research in Brazil]*. Lavras:: Editora UFLA, 2010.

Smith, S. E. and Read, D. J. *Mycorrhizal symbiosis*. Academic Press, 2008.

Sokol, Noah W., Eric Slessarev, Gianna L. Marschmann, Alexa Nicolas, Steven J. Blazewicz, Eoin L. Brodie, Mary K. Firestone, Megan M. Foley, Rachel Hestrin, Bruce A. Hungate, Benjamin J. Koch, Bram W. Stone, Matthew B. Sullivan, Olivier Zablocki, LLNL Soil Microbiome Consortium & Jennifer Pett-Ridge. 2022. Life and death in the soil microbiome: how ecological processes influence biogeochemistry. *Nature Reviews Microbiology* 20:415-430.

Song, Y. N., Zhang, F. S., Marschner, P., Fan, F. L., Gao, H. M., Bao, X. G., Sun, J. H., Li, L. 2007. Effect of intercropping on crop yield and chemical and microbiological properties in rhizosphere of wheat (Triticum aestivum L.), maize (Zea mays L.), and faba bean (Vicia faba L.). *Biology and Fertility of Soils* 43:565-574.

Starr, E. P., Shi, S., Blazewicz, S. J., Probst, A. J., Herman, D. J., Firestone, M. K., and Banfield, J. F. 2018. Stable isotope informed genome-resolved metagenomics reveals that saccharibacteria utilize microbially-processed plant-derived carbon. *Microbiome* 6:1-12.

Staton, T., Breeze, T. D., Walters, R. J., Smith, J., Girling, R. D. 2022. Productivity, biodiversity trade-offs, and farm income in an agroforestry versus an arable system. *Ecological Economics* 191:107214.

Sun, B., Gao, Y., Wu, X. Ma, H., Zheng, C., Wang, X., Zhang, H., Li, Z., Yang, H. 2020. The relative contributions of pH, organic anions, and phosphatase to rhizosphere soil phosphorus mobilization and crop phosphorus uptake in maize/alfalfa polyculture. *Plant and Soil* 447:117-133.

Suprapta, D. N. 2012. Potential of microbial antagonists as biocontrol agents against plant fungal pathogens. *Journal of International Society for Southeast Asian Agricultural Sciences* 18:1-8.

Suzuki, Y. 2020. Advent of a new sequencing era: long-read and on-site sequencing. *Journal of Human Genetics* 65:10038.

Tahir, H. A. S., Gu, Q., Wu, H., Raza, W., Hanif, A., Wu, L., Colman, M. V., and Gao, X. 2017. Plant growth promotion by volatile organic compounds produced by Bacillus subtilis SYST2. *Frontiers in Microbiology* 8:1-11.

Trivedi, P., Delgado-Baquerizo, M., Trivedi, C., Hamonts, K., Anderson, I. C., Singh, B. K. 2017. Keystone microbial taxa regulate the invasion of a fungal pathogen in agro-ecosystems. *Soil Biology and Biochemistry* 111:10-14.

Uren, N. C. Types, amounts, and possible functions of compounds released into the rhizosphere by soil-grown plants. In: Pinton, R.; Varani, Z.; Nannipieri, P. (Eds.). *The rhizosphere: biochemistry and organic substances at the soil-plant interface*. CRC Press, 2009.

Van Elsas, J. D., Chiurazzi, M., Mallon, C. A., Elhottova, D., Kristufek, V., Salles, J. F. 2012. Microbial diversity determines the invasion of soil by a bacterial pathogen. *Proceedings of the National Academy of Sciences* 109:1159-1164.

van Veelen, A., Tourell, M. C., Koebernick, N., Pileio, G., Roose, T. 2018. Correlative visualization of root mucilage degradation using X-ray CT and MRI. *Frontiers in Environmental Science* 6:1-9.

Vannier, N., Agler, M., Hacquard, S. 2019. Microbiota-mediated disease resistance in plants. *PLOS Pathogens* 15:e1007740.

Vazquez, M. M., César, S., Azcón, R., Barea, J. M. 2000. Interactions between arbuscular mycorrhizal fungi and other microbial inoculants (Azospirillum, Pseudomonas, Trichoderma) and their effects on microbial population and enzyme activities in the rhizosphere of maize plants. *Applied Soil Ecology* 15:261-272.

Venturi, V., Keel, C. 2016. Signaling in the Rhizosphere. *Trends in Plant Science* 21:187-198.

Verhagen, B. W. M., Glazebrook, J., Zhu, T., Chang, Hur-Song., van Loon, L. C., and Pieterse, C. M. J. 2004. The transcriptome of rhizobacteria-induced systemic resistance in Arabidopsis. *Molecular Plant-Microbe Interactions* 17:895-908.

Vetterlein, D., Carminati, A., Kögel-Knabner, I., Bienert, G. P., Smalla, K., Oburger, E., Schnepf, A., Banitz, T., Tarkka, M. T., and Schlüter, S. 2020. Rhizosphere spatiotemporal organization–a key to rhizosphere functions. *Frontiers in Agronomy* 2:1-22.

Vives-Peris, V., Ollas, C., Gómez-Cadenas, A., Pérez-Clemente, R. 2020. Root exudates: from plant to rhizosphere and beyond. *Plant Cell Reports* 39:3-17.

Vries, F. T., Williams, A., Stringer, F., Willcocks, R., McEwing, R., Langridge, H., Straathof, A. L. 2019. Changes in root-exudate-induced respiration reveal a novel mechanism through which drought affects ecosystem carbon cycling. *New Phytologist* 224:132-145.

Walker, T. S.; Bais, H. P.; Grotewold, E.; Vivanco, J. M. 2003. Root exudation and rhizosphere biology. *Plant Physiology* 132:44-51.

Wang, X., Tang, C., Severi, J., Butterly, C. R., Baldock, J. A. 2016. Rhizosphere priming effect on soil organic carbon decomposition under plant species differing in soil acidification and root exudation. *New Phytologist* 211:864-873.

Wang, X., Yang, Y., Pei, K., Zhou, J., Peixoto, L., Gunina, A., Zeng, Z., Zang, H., Rasmussen, J., Kuzyakov, Y. 2021. Nitrogen rhizodeposition by legumes and its fate in agroecosystems: A field study and literature review. *Land Degradation & Development* 32:410-419.

Wang, F., Zhang, X., Wei, M., Wang, Y., Liang, Z., & Xia, P. 2022. Appropriate crop rotation alleviates continuous cropping barriers by changing rhizosphere microorganisms in Panax notoginseng. *Rhizosphere* 23:100568.

Watson, J., and Crick, F. 1953. Molecular structure of nucleic acids. *Nature* 171 (4356): 737:738.

Wei, Z., Gu, Y., Friman, V-P., Kowalchuk, G. A., Xu, Y., Shen, Q., Jousset, A. 2019. Initial soil microbiome composition and functioning predetermine future plant health. *Science Advances* 5:eaaw0759.

Wei, Z., Hu, J., Gu, Y., Yin, S., Xu, Y., Jousset, A., Shen, Q., Friman, V. P. 2018. *Ralstonia solanacearum* pathogen disrupts bacterial rhizosphere microbiome during an invasion. *Soil Biology and Biochemistry* 118:8-17.

Wei, Z., Yang, T., Friman, V. P., Xu, Y., Shen, Q., Jousset, A. 2015. Trophic network architecture of root-associated bacterial communities determines pathogen invasion and plant health. *Nature Communications* 6:8413.

Wen, J., Li, Z., Luo, N., Huang, M., Yang, R., Zeng, G. 2018. Investigating organic matter properties affecting the binding behavior of heavy metals in the rhizosphere of wetlands. *Ecotoxicology and Environmental Safety* 162:184-191.

Whalley, W. R., Riseley, B., Leeds-Harrison, P. B., Bird, N. R. A., Leech, P. K., Adderley, W. P. 2005. Structural differences between bulk and rhizosphere soil. *European Journal of Soil Science* 56:353-360.

Wille, L., Messmer, M. M., Studer, B. Hohmann, P. 2019. Insights to plant-microbe interactions provide opportunities to improve resistance breeding against root diseases in grain legumes. *Plant, Cell & Environment* 42:20-40.

Wilschut, R. A., Geisen, S. 2021. Nematodes as drivers of plant performance in natural systems. *Trends in Plant Science* 26:237-247.

Woese, C. R., Fox, G. E. 1977. Phylogenetic structure of the prokaryotic domain: the primary kingdoms. *Proceedings of the National Academy of Sciences of the United States of America* 74:5088-5090.

Wright, S. F., Franke-Snyder, M., Morton, J. B., Upadhyaya, A. 1996. Time-course study and partial characterization of a protein on hyphae of arbuscular mycorrhizal fungi during active colonization of roots. *Plant and Soil* 181:193-203.

Wright, S. F., Upadhyaya, A. 1998. A survey of soils for aggregate stability and glomalin, a glycoprotein produced by hyphae of arbuscular mycorrhizal fungi. *Plant and Soil* 198:97-107.

Wu, H., Qin, X., Wang, J., Wu, L., Chen, J., Fan, J., Zheng, L., Tangtai, H., Arafat, Y., Lin, W., Luo, X. 2019. Rhizosphere responses to environmental conditions in *Radix pseudostellariae* under continuous monoculture regimes. *Agriculture, Ecosystems & Environment* 1(270):19-31.

Yin, R., Hao, Z., Qu, L., Wu H., Du, X., Yuan, X., Zhang, X., Chen, B. 2022. Mycorrhizal symbiosis and water condition affect ozone sensitivity of Medicago sativa L. by mediating stomatal conductance. *Environmental and Experimental Botany* 202(18):105037.

York, L. M., Carminati, A., Mooney, S. J., Ritz, K., and Bennett, M. J. 2016. The holistic rhizosphere: Integrating zones, processes, and semantics in the soil influenced by roots. *Journal of Experimental Botany* 67(12):3629-3643.

Zarebanadkouki, M., Ahmed, M. A., Carminati, A. 2016. Hydraulic conductivity of the root-soil interface of lupin in sandy soil after drying and rewetting. *Plant and Soil* 398:267-280.

Zarebanadkouki, M., Ahmed, M. A., Hedwig, C., Benard, P., Kostka, S. J., Kastner, A., Carminati, A. 2018. Rhizosphere hydrophobicity limits root water uptake after drying and subsequent rewetting. *Plant and Soil* 428:265-277.

Zarebanadkouki, M., Kim, Y. X., Moradi, A. B., Vogel, H. J., Kaestner, A., Carminati, A. 2012. Quantification and modeling of local root water uptake using neutron radiography and deuterated water. *Vadose Zone Journal*, 11(3).

Zarebanadkouki, M., Kroener, E., Kaestner, A., Carminati, A. 2014. Visualization of root water uptake: Quantification of deuterated water transport in roots using neutron radiography and numerical modeling. *Plant Physiology* 166:487-499.

Zeng, M., Hause, B., Van Dam, N. M., Uthe, H., Hoffmann, P., Krajinski, F., and Martínez-Medina, A. 2022. The mycorrhizal symbiosis alters the plant defence strategy in a model legume plant. *Plant Cell and Environment* 45(12):1–17.

Zhang, M. M., Wang, N., Hu, Y. B., Sun, G. Y. 2018. Changes in soil physicochemical properties and soil bacterial community in mulberry (*Morus alba* L.)/alfalfa (*Medicago sativa* L.) intercropping system. *Microbiology open* 7(2):e00555.

Zhang, M., Chai, L., Huang, M., Jia, W., Guo, J., Huang, Y. 2020. Deciphering the archaeal communities in tree rhizosphere of the qinghai-tibetan plateau. *BMC Microbiology* 20(1):1-13.

Zhang, X., Baars, O., and Morel, F. M. M. 2019. Genetic, structural, and functional diversity of low and high-affinity siderophores in strains of nitrogen fixing: *Azotobacter chroococcum*. *Metallomics* 11(1):201-12.

Zhong, X. L., Li, J. T., Naveed, M., Raffan, A., Hallett, P. D. 2021. A laboratory study to disentangle hydrological, mechanical and structural mechanisms of soil stabilization by plant mucilage between eroding and depositional zones of a slope. *European Journal of Soil Science* 72:125-140.

Zhou, D., Feng, H., Schuelke, T., Santiago, A., Zhang, Q., Zhang, J., Luo, C., Wei, L. 2019. Rhizosphere microbiomes from root knot nematode non-infested plants suppress nematode infection. *Microbial Ecology* 78(2):470-81.

Zhou, J., Shao, G., Kumar, A. Shi, L., Kuzyakov, Y., Pausch, J. 2022. Carbon fluxes within tree-crop-grass agroforestry system: 13C field labeling and tracing. *Biology and Fertility of Soils* 58:733-743.

Index

N

O

P

About the Editors

Rodrigo Nogueira de Sousa, PhD
Researcher,
Soil Science Department,
University of São Paulo, "Luiz de Queiroz"
College of Agriculture
Piracicaba, SP, Brazil

Rodrigo Nogueira de Sousa obtained an undergraduate degree in Agronomic Engineering and a master's in Soil Science and Plant Nutrition from the Federal University of Viçosa (UFV), Brazil, in 2016 and 2018, respectively. He obtained a Ph.D. in Soil Science and Plant Nutrition from the University of São Paulo (USP). From 2014 to 2015, he studied at North Carolina Agricultural and Technical State University, USA. During his doctorate program, he studied the feasibility of alternative sources of phosphate fertilizers in tropical soils.

Thais Nascimento Pessoa, PhD
PostDoc
University of São Paulo, "Luiz de Queiroz"
College of Agriculture
Piracicaba, SP, Brazil

Thais Nascimento Pessoa obtained her Bachelor's degree in Agronomy from the State University of Southwest Bahia (UESB) between 2009 and 2014, and her Master's degree in Agronomy, Soil and Plant Nutrition from the Federal University of Viçosa (UFV) between 2014 and 2016. Thais is also a Specialist and Ph.D. in Soil Science and Plant Nutrition, having obtained her degrees from the "Luiz de Queiroz" College of Agriculture/University of São Paulo (ESALQ/USP) between 2017 and 2019 and 2016 and 2020, respectively. During her Ph.D. studies, Thais spent some time at the Laboratory of Mechanics, Hydrology, and Image Analysis at the University of Kassel, Witzenhausen, Germany.

Kassio Ferreira Mendes, PhD
Professor
Department of Agronomy
Federal University of Viçosa,
Viçosa, MG, Brazil

Kassio Ferreira Mendes is a Professor of Biology and Integrated Management of Weeds, Department of Agronomy, Federal University of Viçosa, Brazil. He graduated from the State University of Mato Grosso in 2011 and received a Master's in Agronomy (Crop Science) from the Federal University of Viçosa in 2013. He obtained a doctorate and post-doctorate degrees in Sciences - Nuclear Energy in Agriculture from the Center of Nuclear Energy in Agriculture, University of São Paulo, Brazil, in 2017 and 2019, respectively. He completed research fellowships at the University of Minnesota, USA.